Fatos Xhafa, Santi Caballé, Ajith Abraham, Thanasis Daradoumis, and Angel Alejandro Juan Perez (Eds.)

Computational Intelligence for Technology Enhanced Learning

Studies in Computational Intelligence, Volume 273

Editor-in-Chief
Prof. Janusz Kacprzyk
Systems Research Institute
Polish Academy of Sciences
ul. Newelska 6
01-447 Warsaw
Poland
E-mail: kacprzyk@ibspan.waw.pl

Further volumes of this series can be found on our homepage: springer.com

Vol. 253. Roger Lee and Naohiro Ishii (Eds.)
Software Engineering Research, Management and Applications 2009, 2009
ISBN 978-3-642-05440-2

Vol. 254. Kyandoghere Kyamakya, Wolfgang A. Halang, Herwig Unger, Jean Chamberlain Chedjou, Nikolai F. Rulkov, and Zhong Li (Eds.)
Recent Advances in Nonlinear Dynamics and Synchronization, 2009
ISBN 978-3-642-04226-3

Vol. 255. Catarina Silva and Bernardete Ribeiro
Inductive Inference for Large Scale Text Classification, 2009
ISBN 978-3-642-04532-5

Vol. 256. Patricia Melin, Janusz Kacprzyk, and Witold Pedrycz (Eds.)
Bio-inspired Hybrid Intelligent Systems for Image Analysis and Pattern Recognition, 2009
ISBN 978-3-642-04515-8

Vol. 257. Oscar Castillo, Witold Pedrycz, and Janusz Kacprzyk (Eds.)
Evolutionary Design of Intelligent Systems in Modeling, Simulation and Control, 2009
ISBN 978-3-642-04513-4

Vol. 258. Leonardo Franco, David A. Elizondo, and José M. Jerez (Eds.)
Constructive Neural Networks, 2009
ISBN 978-3-642-04511-0

Vol. 259. Kasthurirangan Gopalakrishnan, Halil Ceylan, and Nii O. Attoh-Okine (Eds.)
Intelligent and Soft Computing in Infrastructure Systems Engineering, 2009
ISBN 978-3-642-04585-1

Vol. 260. Edward Szczerbicki and Ngoc Thanh Nguyen (Eds.)
Smart Information and Knowledge Management, 2009
ISBN 978-3-642-04583-7

Vol. 261. Nadia Nedjah, Leandro dos Santos Coelho, and Luiza de Macedo de Mourelle (Eds.)
Multi-Objective Swarm Intelligent Systems, 2009
ISBN 978-3-642-05164-7

Vol. 262. Jacek Koronacki, Zbigniew W. Ras, Slawomir T. Wierzchon, and Janusz Kacprzyk (Eds.)
Advances in Machine Learning I, 2009
ISBN 978-3-642-05176-0

Vol. 263. Jacek Koronacki, Zbigniew W. Ras, Slawomir T. Wierzchon, and Janusz Kacprzyk (Eds.)
Advances in Machine Learning II, 2009
ISBN 978-3-642-05178-4

Vol. 264. Olivier Sigaud and Jan Peters (Eds.)
From Motor Learning to Interaction Learning in Robots, 2009
ISBN 978-3-642-05180-7

Vol. 265. Zbigniew W. Ras and Li-Shiang Tsay (Eds.)
Advances in Intelligent Information Systems, 2009
ISBN 978-3-642-05182-1

Vol. 266. Akitoshi Hanazawa, Tsutom Miki, and Keiichi Horio (Eds.)
Brain-Inspired Information Technology, 2009
ISBN 978-3-642-04024-5

Vol. 267. Ivan Zelinka, Sergej Celikovský, Hendrik Richter, and Guanrong Chen (Eds.)
Evolutionary Algorithms and Chaotic Systems, 2009
ISBN 978-3-642-10706-1

Vol. 268. Johann M.Ph. Schumann and Yan Liu (Eds.)
Applications of Neural Networks in High Assurance Systems, 2009
ISBN 978-3-642-10689-7

Vol. 269. Francisco Fernández de de Vega and Erick Cantú-Paz (Eds.)
Parallel and Distributed Computational Intelligence, 2009
ISBN 978-3-642-10674-3

Vol. 270. Zong Woo Geem
Recent Advances In Harmony Search Algorithm, 2009
ISBN 978-3-642-04316-1

Vol. 271. Janusz Kacprzyk, Frederick E. Petry, and Adnan Yazici (Eds.)
Uncertainty Approaches for Spatial Data Modeling and Processing, 2009
ISBN 978-3-642-10662-0

Vol. 272. Carlos A. Coello Coello, Clarisse Dhaenens, and Laetitia Jourdan (Eds.)
Advances in Multi-Objective Nature Inspired Computing, 2009
ISBN 978-3-642-11217-1

Vol. 273. Fatos Xhafa, Santi Caballé, Ajith Abraham, Thanasis Daradoumis, and Angel Alejandro Juan Perez (Eds.)
Computational Intelligence for Technology Enhanced Learning, 2010
ISBN 978-3-642-11223-2

Fatos Xhafa, Santi Caballé, Ajith Abraham,
Thanasis Daradoumis, and
Angel Alejandro Juan Perez (Eds.)

Computational Intelligence for Technology Enhanced Learning

Springer

Dr. Fatos Xhafa
Department of Languages and
Informatics Systems
Polytechnic University of Catalonia
Campus Nord, Ed. Omega, C/Jordi Girona 1-3
08034 Barcelona, Spain
E-mail: fatos@lsi.upc.edu

Dr. Thanasis Daradoumis
Open University of Catalonia
Department of Computer Sciences
Multimedia and Telecommunications
Rbla. Poblenou. 156
08018 Barcelona, Spain
E-mail: adaradoumis@uoc.edu

Dr. Santi Caballé
Open University of Catalonia
Department of Computer Sciences
Multimedia and Telecommunications
Rbla. Poblenou. 156
08018 Barcelona, Spain
E-mail: scaballe@uoc.edu

Dr. Angel Alejandro Juan Perez
Open University of Catalonia
Department of Information Sciences
Av. Tibidabo 39-43
08035 Barcelona, Spain
E-mail: ajuanp@uoc.edu

Dr. Ajith Abraham
Machine Intelligence Research Labs
(MIR Labs)
Scientific Network for Innovation
and Research Excellence
P.O. Box 2259
Auburn, Washington 98071-2259, USA
E-mail: ajith.abraham@ieee.org

ISBN 978-3-642-11223-2 e-ISBN 978-3-642-11224-9

DOI 10.1007/978-3-642-11224-9

Studies in Computational Intelligence ISSN 1860-949X

Library of Congress Control Number: 2009943589

Typeset & Cover Design: Scientific Publishing Services Pvt. Ltd., Chennai, India.

Printed in acid-free paper

9 8 7 6 5 4 3 2 1

springer.com

Preface

In this book is presented up-to-date research on computational intelligence oriented to support technology-enhanced learning systems and processes. Regarding the topics of online work and learning, the book aims at providing both instructors and researchers with new approaches for effective and efficient means to assist e-learning participants in enforcing and improving their online learning and working activities –both at individual and group levels. Moreover, the book appeals for providing software developers and researchers in the field of online learning systems with fresh and innovative ideas that allow them to extend current capabilities and functionalities of e-learning platforms. The goal here is to make an efficient use of these e-learning platforms in a distributed environment where learning design and material producers, service providers, and users –either instructors, learners, or academic coordinators– share similar learning and work experiences.

Despite the considerable progress that has been made in recent years in the field of computational intelligence for technology enhanced learning, there are still plenty of issues to investigate on how to employ the emergent computational technologies to fully support online learning and teaching activities. To this end, the chapters in this book give special emphasis on applications of computational-intelligence approaches to a variety of research topics, among others: personalization of learning, mobile learning, adaptive learning, interactive digital TV learning, collaborative learning, web-based tutoring systems, interaction data analysis and mining, or intelligent skill development. Therefore, this book covers the needs and interests of a wide range of readers, giving them the opportunity to deepen further on these issues and also to extend their knowledge to areas other than the ones they are used to work with. Moreover, the merge of all these synergies represents an attractive challenge that will yield systems capable of providing more effective answers on how to improve and enhance the on-line learning and teaching experiences.

Among the many features highlighted in the book, which provide a significant support to the design and development of computational-intelligence e-learning systems and models, we could distinguish the following ones by chapter:

Intelligent Techniques in Personalization of Learning in e-Learning Systems: This chapter contains an overview of intelligent techniques that can be applied in different stages of e-learning systems to achieve personalization, including:

clustering methods, classification methods, rule based systems, etc. It describes examples of their application to various e-learning platforms to create profiles of learners and to define personalized learning paths. The chapter also includes a critical discussion of the existing approaches and suggests possible research lines in this field.

Fuzzy ECA Rules for Pervasive Decision-Centric Personalized Mobile Learning: This chapter addresses personalization in intelligent context-aware information systems. The chapter introduces a personalized mobile learning system as an information system related to the educational domain. Since personalization requires both identification and selection of individuals, the chapter proposes the use of individual profiles and introduces Fuzzy Event: Condition: Action (FECA) rules as an effective approach that allows applying computational intelligence to personalized mobile learning systems. The proposed FECA rules algorithm is introduced and evaluated. The chapter also includes some suggestions of future research lines in this area.

Developing an Adaptive Learning Based Tourism Information System Using Ant Colony Metaphor: This chapter proposes an adaptive-learning algorithm, which is bio-inspired, in the context of Travel Information Systems. Adaptive learning has two features, e.g. diversity and interactivity across the group members. The chapter proposes a model which reveals that intelligent-learning technologies can make use of group's behavior to evaluate the current learning state and, consequently, to recommend the next best move or selection. The final goal of this learning process is to optimize the different available service and product pattern across a heterogeneous group of tourists, which is attained by simulating the pheromone deposition and update mechanism of ants.

Intelligent and Interactive Web-based Tutoring System in Engineering Education: Reviews, Perspectives and Development: During the last years, a large number of Web-based intelligent tutoring systems have been developed and implemented worldwide. This chapter argues that, in despite of this fact, there is still a lack of relevant comprehensive research concerning the efficiency of these systems in the context of engineering education. Therefore, the chapter outlines and discusses important issues related to the development of Web-based intelligent tutoring systems in the engineering education arena. Also, a case study regarding the development of a Web-based computer-assisted laboratory for electrical engineers is presented and discussed.

Granular Mining of Student's Learning Behavior in Learning Management System Using Rough Set Technique: Users' behaviors and interactions in learning management systems can be intelligently examined in order to analyzing students' learning styles. Such behavioral and interaction patterns include the way the user navigate, the links he/she selects among the ones provided by the system, the most popular learning materials, and the usage level of each system tool. In this chapter, a model to analyze students' preferences in online learning systems is proposed. The model is based on the integrated Felder-Silverman learning style model. An example of the proposed methodology, based on the use of Moodle, is also introduced. The chapter also discusses how the students' preferences are consistent with the characteristics of the learning styles described by the Felder-Silverman model.

Personalised hybrid approaches based on ontologies and folksonomies to support T-learning 2.0: The aim of this chapter is to export the technology of collaborative tagging to the field of learning through Interactive Digital TV (IDTV) (t-learning). This technology has become a popular practice to annotate resources on the web, and it is now reaching e-learning initiatives. Previous research describes learning experiences for t-learning based on combining TV programmes and learning elements in order to lure viewers into education and make these experiences more entertaining. One the one hand, the current approach reasons over ontologies for the combination of the different elements, and it tries to take into account the user's point of view towards the contents. On the other hand, it goes a step further and presents a proposal that includes collaborative tagging techniques, complementing ontologies with folksonomies to establish the relationships between the contents linked to create the learning experiences.

Computational intelligence infrastructures to support complex e-learning systems: This chapter is based on the fact that modern on-line learning environments have to enable and scale the involvement of increasing large number of single/group participants who can be geographically distributed, and need to transparently share a huge variety of both software and hardware distributed learning resources. As a result, collaborative learning applications are to overcome important non-functional requirements arisen in distributed environments, such as scalability, flexibility, availability, interoperability, and integration of different, heterogeneous, and legacy collaborative learning systems. Therefore, e-Learning applications need to be developed in a way that overcome these demanding requirements as well as to provide educational organizations with fast, flexible and effective solutions for the enhancement and improvement of the learning performance and outcomes. To this end, this chapter presents, evaluates and validates an innovative engineering software technique that combines the Generic Programming paradigm and Service-Oriented Architectures to construct flexible, distributed, scalable, interoperable and robust applications as much effectively and timely as possible, as key aspects to address the demanding and changing requirements in the current e-Learning domain.

Use of Artificial Intelligence skills to build role-playing platforms for negotiation training: In recent decades, a number of trainers have used role-playing games to teach negotiation skills. This chapter presents a teaching methodology that makes possible to conduct this kind of approaches in a virtual environment. The teaching methodologies exploits a specially-developed technology platform that allows a small community of players to communicate interacts and play online in order to acquire basic notions and rules about negotiation and then shows how to apply this knowledge. The work presents initial results of investigating Artificial Intelligence techniques that can be used to evaluate possibilities of implementation of computer-controlled "artificial players" that embody some intelligent behaviour.

Computational intelligence methods for data analysis and mining of eLearning activities: This chapter investigates the development and use of data mining and computational intelligent approaches to enhance the effectiveness of web-based education and provide better learning environments. The development of such intelligent technologies may not only contribute to the growth of e-learning as an important education method but also enable learners to participate in 'any time, any place' personalized training. This work consists, in its first part, of a survey of the applications of data mining and computational intelligence in web based education during 2004–2009, while in the second part it presents a case study that aims at analyzing students' activities performed in a Learning Management System.

Advanced learning technology systems in mathematics education: Mathematics education is a very active field of research. Among the many issues raised, there is a special interest in the development of professional skills for using effective computational and modelling tools in solving real world problems. This chapter investigates the impact that technologically advanced learning technologies may have on mathematics education. It also discusses how web based approaches might suggest new paradigms of mathematical learning aimed to face the new educational challenges of the modern advanced ICT and Network Society in an effective way.

All in all, Computational Intelligence for Technology Enhanced Learning is a major research theme in the Computational Intelligence and e-Learning research community. It comprises a variety of research topics that focus on developing models and systems that are more powerful and also more adaptable to the learning process, thus providing better answers to the paradigmatic principles of on-line learning and work. The chapters collected in this book provide new insights, findings and approaches both on the analysis and the development of more powerful e-learning settings and environments. Researchers will find in this book the latest trends in these research topics. On the other hand, academics will find practical insights on how to use conceptual and experimental approaches in their daily tasks. Finally, developers from e-learning community can be inspired and put in practice the proposed models in order to evaluate them for the specific purposes of their own work and context.

We hope the readers of this book will find it a valuable resource in their research, development and educational activities in online environments.

Acknowledgment: We would like to thank the authors of the chapters and also the referees for their invaluable collaboration and prompt responses to our enquiries, which enabled completion of this book on time. We gratefully acknowledge the support and encouragement received from Prof. Janusz Kacprzyk, the editor in chief of Springer series *"Studies in Computational Intelligence"*, and Ms Heather King for the great assistance during the editorial work.

Fatos Xhafa's research work partially completed at Birkbeck, University of London, on Leave from Technical University of Catalonia (Barcelona, Spain). His research is supported by a grant from the General Secretariat of Universities of the Ministry of Education, Spain.

October 2009

Fatos Xhafa
Santi Caballé
Ajith Abraham
Thanasis Daradoumis
Angel Alejandro Juan Perez

Contents

List of Contributors

Ajith Abraham
Machine Intelligence Research Labs (MIR Labs)
P.O. Box 2259, Washington, USA

Shawki Al-Dubaee
Technical University of Ostrava, Department of Computer Science
Czech Republic

Nor Bahiah Hj Ahmad
Faculty of Computer Science and Information Systems
Universiti Teknologi Malaysia
Malaysia

Arash Bahrammirzaee
Images, Signals and Intelligent Systems Laboratory, PARIS 12 – Val de Marne University, France

Soumya Banerjee
Department of Computer Science and Engineering, Birla Institute of Technology, India

Santi Caballé
Department of Computer Science
Open University of Catalonia, Spain

Monica Chis
Siemens Program and System Engineering Romania

Amine Chohra
Images, Signals and Intelligent Systems Laboratory, PARIS 12 – Val de Marne University, France

G.S. Dangayach
Department of Mechanical Engineering, Malviya National Institute of Technology, India

Rebeca P. Díaz-Redondo
Department of Telematic Engineering
University of Vigo, Spain

Pavla Drazdilova
Technical University of Ostrava, Department of Computer Science
Czech Republic

Ana Fernández-Vilas
Department of Telematic Engineering
University of Vigo, Spain

Bin Hu
Department of Computing, Birmingham City University, Birmingham, UK

Mike Jackson
Department of Computing, Birmingham City University, Birmingham, UK

Dalel Kanzari
Images, Signals and Intelligent Systems Laboratory, PARIS 12 – Val de Marne University, France

Halina Kwasnicka
Wroclaw University of Technology, Institute of Applied Informatics, Poland

Kurosh Madani
Images, Signals and Intelligent Systems Laboratory PARIS 12 – Val de Marne University, France

Jan Martinovic
Technical University of Ostrava
Department of Computer Science
Czech Republic

Philip Moore
Department of Computing, Birmingham City University, Birmingham, UK

Gamila Obadi
Technical University of Ostrava
Department of Computer Science
Czech Republic

Arun S. Patil
CQUniversity Mackay Campus
Australia

José J. Pazos-Arias
Department of Telematic Engineering
University of Vigo, Spain

Katerina Slaninova
Technical University of Ostrava
Department of Computer Science
Czech Republic

Mariusz Paradowski
Wroclaw University of Technology
Institute of Applied Informatics, Poland

Loredana Randazzo
Department of Food Science
University of Naples Federico II, Italy

Marta Rey-López
Department of Telematic Engineering
University of Vigo, Spain

Siti Mariyam Shamsuddin
Faculty of Computer Science and Information Systems
Universiti Teknologi Malaysia
Malaysia

Urszula Markowska-Kaczmar
Wroclaw University of Technology, Institute of Applied Informatics, Wroclaw, Poland

Antonello Santini
Department of Food Science, University of Naples Federico II, Portici - Napoli (Italy)
asantini@unina.it

Vaclav Snasel
FEECS, VSB - Technical University of Ostrava, Department of Computer Science, Czech Republic

Valeria Marina Monetti
Department of Agricultural Engineering, University of Naples Federico II, Italy

Gerardo Toraldo
High Performance Computing and Networking Institute
Italian National Research Council
Napoli, Italy

Fatos Xhafa
Department of Computer Science and Information Systems, Birkbeck University of London, UK
(*On Leave from Technical University of Catalonia, Spain*)

Intelligent Techniques in Personalization of Learning in e-Learning Systems

Urszula Markowska-Kaczmar, Halina Kwasnicka, and Mariusz Paradowski

Wroclaw University of Technology, Institute of Applied Informatics, Wyb. Wyspianskiego 27, 50-370 Wroclaw, Poland

Abstract. This chapter contains an overview of intelligent techniques that can be applied in different stages of e-learning systems to achieve personalization. It describes examples of their application to various e-learning platforms to create profiles of learners and to define learning path. The typical approach to obtain learner's profile is the usage one of the clustering methods, such as: the simple k-means, Self Organizing Map, hierarchical clustering or fuzzy clustering. Classification methods like: C4.5 or C.5, k-Nearest Neighbor and Naive Bayes are also useful, but they need to define classes and training patterns by an expert. In contrary, clustering is unsupervised learning method and the categories are discovered by the method itself. The recommending system is responsible for proposing individual learning path for each learner. The most popular approach is an application of the Aprori method which searches for association rules. However, it seems that it is rather inefficient method when the number of data to process is huge. Other methods and models that can be useful for knowledge representation are also discussed. Recommending systems are mainly built as a knowledge based. Most of them are implemented as rule based systems. An interesting approach implementing cased based reasoning paradigm to recommend learning path is described as well. The end of the chapter contains a critical discussion of existing solutions and suggests possible research in this field.

1 Introduction

E-learning is the method of a remote realization of didactic process with the usage of computer technology that enables a direct contact between a teacher and a learner in real time. In this training process different techniques based on text, image, video and sound are used. From a technical point of view, e-learning system is composed of a user application that provides user communication with e-learning system, e-learning application servers which aim is to interpret user queries and sending HTML code to a client, e-learning content servers that give an access to files with training materials and database server that stores information about servers names and names of users, authorizations to manage the users, didactic contents and trainings. These servers contain metadata connected with contents and a progress of learner training.

In this chapter, however we will focus on another aspect of e-learning system. It is widely known that the method of learning and its pace should be fitted to the knowledge level, the way of knowledge absorption and a learner personality.

F. Xhafa et al. (Eds.): Computational Intelligence for Tech. Enhanced Learning, SCI 273, pp. 1–23.
springerlink.com

During a classical learning process a human – teacher observes and evaluates a skill of a given pupil and is able to adapt learning program and pace of training individually to the student or to the group of students. E-learning systems should go in the same direction.

That is why new e-learning systems are built as adaptive ones. They are capable to monitor the activities of their users; interpreting these on the basis of domain-specific models; to infer about user's requirements and preferences and finally, they act upon the available knowledge on its users and dynamically facilitate the learning process. Because the system behavior adapts to a person, this kind of adaptation is also called personalization.

Generally speaking, personalization means learner-specific strategies to address individual needs and expectations to support and to promote individual learning success. Personalization consists in establishing individual path of learner training on the basis on its personality and training progress. This process may refer to an individual content selection planning from existing repositories, as well as to a dynamic change of repositories contents. Because this process is hardly to algorithmize in many systems intelligent techniques are applied.

On the basis of our study it seems that intelligent techniques offer much in this area. We hope that knowledge of the existing solutions based on this overview can help potential new developers in the successful application of these intelligent methods in their projects. The mission of this chapter is a survey of existing solutions in different e-learning platforms to the problem of learning personalization with the main stress put on the usability evaluation of different intelligent techniques. The representation of knowledge will be also considered.

2 Related Work

Good surveys of existing approaches until the year 2005 with the usage of intelligent methods to implement e-learning systems can be found in [39] and [5]. They contain an overview of many data mining techniques [19] useful for this purpose. Data mining is the process of extracting patterns from data. It involves four classes of tasks: classification, clustering, regression and association rules. The methods to achieve these tasks come from artificial intelligence. *Nearest neighbor*, *Naive Bayes* classifier, *decision trees* and *neural networks* are the examples of common classifiers. They try to arrange the data into predefined groups. Clustering like classification arrange data in categories, but the groups are not predefined, so the algorithm will try to group similar items together. Classification methods are trained in supervised way and clustering methods are trained unsupervised way. Supervised methods to train the model need a help of expert. He/she has to prepare for each input data a label of a class. These pairs create training set for a classifier. In contrary to supervised methods, clustering does not need classes definition within the training set. In the context of personalization both kinds of methods are used for profile creation. For classification a feedforward neural network, decision trees ([9], [13], [14], [27], [43]), simple *k-NN* [7] method or *Naive Bayes* [49] are applied most frequently. *Support Vector Machines* (SVM) which are one of the best classifiers can be also useful [33]. Clustering is realized by the simple *k-means* [37], [55] or *k-medoids* [51], *self organizing* neural network (SOM) [24], *genetic algorithm* [38] or hierarchical clustering methods [43], [47], [51]. These methods produce exclusive clusters of data. Overlapping clusters are obtained by fuzzy clustering methods [48].

Association rules search for relationships between variables (features) so in the context of e-learning systems they are useful for searching for individual learning path, which is form of an advise generated by a recommending system. *Apriori* algorithm and its improvements are applied most commonly for this purpose [23], [26], [30], [36], [37], [46]. Also genetic algorithm can be implemented for this reason The example can be found in [20].

Recommending system can be built as a knowledge based intelligent decision support system where only suggestions for a learners in a form of a learning path are possible, but also as more advanced decision systems– i.e. expert systems. In this case there is a possibility of explanation why a given suggestion (in this case learning path) was proposed [2]. Another paradigm very useful for this purpose is *Case Based Reasoning* (CBR). The example of its application is described in [20].

Intelligent agents are also worth mentioning here. While implemented they usually realize all functions of e-learning systems. A software agent is able undertaking autonomic actions in this environment to achieve a defined goal. An *intelligent agent* should reveal some additional features [1], [24] like: autonomy, adaptation, learning from experience, it should act and collaborate with others, should act transparently and continuously. Examples of agent systems which also realized personalization can be found in [1], [24], [22].

3 Idea of Personalization in e-Learning Systems

Personalization of learning expresses an individual approach to the learner in e-learning systems. It includes all actions designed to match the selected aspects of e-learning course to individual user's needs and opportunities. The main idea of personalization in e-learning systems is presented in Fig.1.

Personality tests and personal data deliver a lot of information about the learner. They refer to the preferred learning way about new things. This is a static view of learner. Dynamic information about the way of the user's learning is acquired from the tracking system. The responsibility of the tracking system is to monitor the behavior of the learner during training. Acquired in this way information describes the results of the current level of knowledge and skills of the student. These both sources of information are used to create a user profile that is, a certain category of learners. Each category is associated with learning path specific for this profile. This path can refer to various aspects of individual learning as for instance content of a course, presentation of the content, navigation or collaboration with the system.

Personalization can be done on-line. In this case the system monitors the learner and system interaction in real time and adapts learning path according to the characteristics of the learner. Reasoning mechanism, ontology for the content of courses and data mining methods for tracking references to a web page are applied in the off-line mode. The system collects learner's data, analyzes them and recommends a teacher changes in the course content.

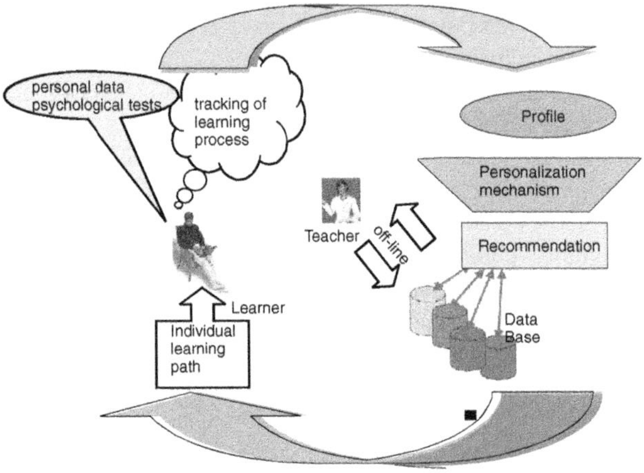

Fig. 1. The idea of learning personalization in e-learning systems

Our presentation of the intelligent methods applied to solve personalization in e-learning systems will be divided into two parts – the first one referring to creation of user's profiles and the second one which describes how these methods are applied to recommend individual path of learning.

4 Intelligent Methods in Creation of User's Profile

Let us recall – typical group of students are called profile or sometimes a student model [14], [16]. The role of student's profile is to automatically discover the user-learner preferences and needs.

4.1 Data Useful in e-Learning Systems for Profile Creation Purpose

Because user related information is fundamental to the personalization of user queries and archive responses we will start an overview from the beginning – from data which influence on results of personalization and can restrict the scope of usable methods. User profiling can be realized as passive (static) or active (dynamic). The former is mostly realized as an automatic procedure delivering data, which are not altered during the learner-system interaction, whereas the latter requires active user involvement. Data acquired in a static way mostly [16] refer to: personal, personality, cognitive, pedagogical and preference data. Personal data describe the biographical information about the student. Personality data come from personality tests. They inform about ability of student concentration, cooperation and relational skills. Cognition tests deliver information about the type of student's cognition. Learning style and the approach to learning are defined in pedagogical data.

Data acquired in a dynamic way contain: the performance data and the learner's (student's) knowledge data. They come from the student-system interaction. Performance data describe the student's current performance in the course sessions. The student's knowledge data describe the knowledge concepts and competences relevant for the current course that the student possesses and must possess until the end of the course.

In most projects data objects represent information items, context objects are associated to data objects and state in which context items are visible. Behavior objects are associated to data objects and include actions, which are performed, whenever some events on the data object occur. Links, concepts, levels of detail, and tours represent different association over objects. These concepts are used within the contextual link server, which groups and references the resources in the web by means of these concepts. In ELENA [11] project data objects are resources, which are annotated by metadata representing the different kinds of attributes described. Resources also can have associated accessibility restrictions.

The data about a learner and its knowledge, acquired by e-learning system, are very important for a personalization purpose. Independently in which way the data are acquired or what was the source of this information they should be recorded and formalized to be usable for further processing in the system. Because of their complexity a semantic representation is needed. That is why ontology and semantic networks are applied. Ontology specifies relevant concepts and the semantics relationships that exist between those concepts in a particular domain. Semantic networks represent knowledge in a form of graph of interconnected nodes. They include concepts represented by nodes and relationships between concepts or propositions, indicated by a connecting line between two concepts.

4.2 Typical Approaches to Profile Creation

In order to create profiles the students are grouped. Again, two approaches can be considered in this case. So called *content based filtering* methods where a student profile is built up by analyzing the content of items that the user has rated in the past and/or user's (learner's) personal information and preferences. The user's profile are constructed by analyzing the responses to a questionnaire, item ratings, or the user's navigation information to infer the user's preferences and/or interests. The shortcomings of this solution is a very shallow analysis of specific kinds of content (text documents, etc.) and the limitation that users can receive only recommendations similar to their earlier experiences.

In contrary to the *content based filtering* the searching for similarity of a group of students in order to define profile is called *collaborative filtering*. If some users have similar behavior, they will be categorized to the same user group. This approach is very popular and we will present it further. It is worth mentioning that collaborative filtering system also has some shortcomings, including that the coverage of item ratings could be very sparse and that it is difficult to provide services for users who have unusual tastes, and the user clustering and classification problems for users with changing and/or evolving preferences [51].

No matter what kind of filtering is applied we can assume that we have the set G of objects o_i (in our case these objects are learners) to each of them a vector $\boldsymbol{v}_i$ is assigned, which is expressed as follows:

$$v_i = [\, v_{i,1}, v_{i,2}, \ldots v_{i,n} \,] . \tag{1}$$

The attributes $v_{i,j}$ included in this vector can be: numerical, categorical or boolean. They can refer to the features describing a learner but they can also represent a sequence of frames of learner's sessions based on a specific time interval.

In order to search group of students with similar characteristics two approaches are applied: *clustering* and *classification*. The first one groups learners with similar characteristics of learning in one category which is called cluster. Unsupervised machine learning methods are used in this case. This means that the classes of users are not known in advance. They are found on the basis of similarity between learners or patterns – as we say in machine learning domain.

The function of the similarity of objects are mapped pair (o_i, o_j) in the set G of a real number $S_{ij}= s(o_i, o_j)$, which determines the similarity of these objects. The function s can be expressed as follows:

$$s : GxG \rightarrow R . \tag{2}$$

The principle of clustering is to maximize the similarity inside an objects group and to minimize the similarity between the object groups. A method for computing similarity between objects mostly based on a distance calculation. Different distance measures can be applied. Most frequently it is the Euclidean distance, but scalar product is also used.

The second possibility to create students profile is to use a classification method [25]. This is a supervised induction method that needs for each pattern $\boldsymbol{v}_i$ a label of a class to which this example belongs. In other word training set consists of pairs – an input pattern (the vector $\boldsymbol{v}_i=[v_{i,1}, v_{i,2}, \ldots v_{i,k}]$) and the label of its class c_i. The classes are defined by experts.

All these supervised and unsupervised methods represent inductive learning. This means that they are trained on the representative patterns and then in the normal phase of a system performance they produce the class or a category of a given pattern (learner). Each learner in the method is described by a feature vector. These data are the basis of the profile creation.

4.3 Survey of Intelligent Methods Applied in Profile Creation

In [47] the learner's profiling scheme consists of two profiling approaches; both of them start on different grounds, but in the last phase their results are combined. In the first one the profiles are constructed by experts. From the point of view of this chapter the second approach is more interesting. It is based on the application of a data clustering technique whose main goal is to identify homogeneous groups of objects based on the values of their attributes.

Because the original set of questions of the e-questionnaires as input space, results into a huge number of unique features to be taken into consideration a hierarchical clustering algorithm was applied, which does not demand the number of clusters as input a priori. Its pseudocode is shown in Fig. 2. Hierarchical clustering of data enables to find groups of data, while gaining knowledge about the relationship between groups of data. These methods therefore offer more information than the flat methods of clustering. Two approaches can be applied in hierarchical clustering. In the *top-down* methods initially, all objects are assigned to one cluster; in next iterations, the cluster is being divided into smaller clusters. This process is repeated until stop criterion is met. In the *bottom-up* methods initially each object constitutes the separate cluster, then clusters are joined in greater clusters. This approach is implemented in [47]. The authors applied Euclidean distance as a distance measure *d* (Fig.2.). Moreover, the interesting is that weighting of features is used because some features are less important than others. The groups obtained in clustering process are then applied to classify the responses of learners to specific parts of the questionnaire.

1. Turn each input element into a singleton, i.e, into a cluster of a single element.
2. For each pair of clusters c1, c2 calculate their distance *d*(c1, c2).
3. Merge the pair of clusters that take the smallest distance.
4. Continue the step 2, until the termination criterion is satisfied.
5. The termination criterion most commonly used is a threshold of the distance value

Fig. 2. The hierarchical clustering algorithm in pseudocode

In the exclusive clustering like in the above case, data are grouped in an exclusive way, so that if a data belong to a definite cluster then it could not be included in another one. On the contrary, the overlapping clustering uses fuzzy sets to cluster data, so that each object may belong to two or more clusters with different degrees of membership. An application of fuzzy hierarchical clustering algorithm w SPERO system is described in [48]. This clustering is applied to only a part of the system data set in question and then refined and extended to the whole data set. The performance of the proposed methodology was compared to the previous static step, using the predefined profile characterizations as label information. In the SPERO system the clustering results demonstrate the clear trend underlying in the system input data: learners are characterized by intermediate skills and expertise of the user.

The advantage of hierarchical clustering is that it does not demand the number of clusters as many other clustering algorithms do, for instance well known *k-means* or *k-medoids*. The pseudocode of *k*-means algorithm is presented in Fig. 3.

1. Select randomly *k* objects (patterns) to be the seeds for the *centroids* of *k* clusters.
2. Assign each pattern to the *centroid* closest to the example, forming in this way *k* exclusive clusters of examples.
3. Calculate new *centroids* of the clusters. For that purpose average all attribute values of the examples belonging to the same cluster (*centroid*).
4. Check if the cluster *centroids* have changed If yes, start again the step 2). If not, cluster detection is finished and all patterns have their cluster memberships defined.

Fig. 3. *k-means* clustering algorithm

In the *k-means* method the mean value of the objects in a cluster, which can be viewed as the cluster centroid, is the center of gravity. The *k-medoids* method uses the *medoid* – the most centrally located object cluster.

The *k-means* algorithm is applied in the AHA! e-learning system presented in [37]. Each of the clusters obtained corresponds to a specific student's model (profile) and is stored in an XML file. In this file the centroid of each cluster is stored. In some sense the centroid represents a typical user of the cluster.

In [51] a combination of *k-medoids* and hierarchical clustering is used. The algorithm is called HBM. It is applied to cluster users based on the time framed navigation sessions. For this purpose the authors define user similarity based on the concept of time-framed navigation sessions. In other words, each two users o_i and o_j are described by their time-framed (TF) navigation sessions S_k, which is a collection of web pages that the users have visited during a session. Similarity of users means the two users have similar navigation behavior during the periods of time intervals. First the similarity of two session records, is defined which is expressed as a ratio of common frames of two records to the sum of frames both of them. Next, similarity of the two time-framed sessions is calculated by choosing a minimum from an average of maximal similarity between records of these session.

Some research is conducted in order to find efficient clustering method. As an example the paper [55] may be mentioned here. Its authors proposed an improvement of *matrix-based clustering* method to group learners. Shortly characterizing this clustering method, the features v_i are weighted and feature vectors are normalized. Then the similarity between each two objects (two feature vectors) is calculated and placed in matrix **SM**. This similarity measure is constructed in this way that the larger its value is, the more features two considered learners have in common. By applying a permutation of similarity matrix columns the best clustering is searched. Strictly, the dividing point *D* along the main diagonal of matrix **SM** is searched. It is implemented by partitioning the permuted matrix into four sub-matrices along its main diagonal. They create two clusters. This matrix partition is recursively applied to the matrices $\mathbf{SM}_{1,1}$ and $\mathbf{SM}_{2,2}$ until the corresponding value of *Agglomerative Strength* of clusters resulting from this division of learners is greater than a threshold *T* or until the number of elements in this cluster is in a given scale. *Agglomerate Strength* is defined as the ratio between the global affinity of the Cluster and the Inside Perfect Cluster with the same size. It means that with higher agglomerate strength, the elements in the cluster are closely similar with each other. In case of this paper the improvement of this method is considered which assumes an application of the *Agglomerate Strength.* The authors compare their results with *k-means* algorithm results, showing that their method has a good efficiency in the given e-learning environment in contrast to the previous matrix-based clustering method and the *k-means* clustering method.

The *Self Organizing Map* (SOM) – this neural network seems to be an interesting method for clustering. The SOM (also known as the Kohonen feature map) constructs a topology preserving mapping from the high-dimensional space created by input feature vectors onto map of neurons in such a way that relative distances between data points are preserved on the map. The neurons, usually form a two-dimensional table, where the location of a neuron in the map carries semantic information. The SOM can thus serve as a clustering tool of high-dimensional data. Because of its typical two-dimensional shape, data can be also easy visualized. This approach was with success applied by us in *Learning Assistant* [24]. Data referring to the style of learning of students are given as inputs to the SOM neural network. These data are acquired from psychological tests fulfilled by learners.

In [34] authors present the system where clustering is applied in order to obtain groups of students with similar characteristics. However they do not give explicitly which method they used for this process. After clustering they applied *Cluster Discrimination* to find key differences among clusters. For instance, a student who prefers teachers to specify topics for essays, is more likely to be arranged in another cluster than a student for whom multimedia materials were the most motivating. Finally students are clustered into three groups with some similar characteristics for all students, like working in teams or passing exams sequentially. The authors do not precise what kind of method they apply for *Cluster Discrimination.* The differences referring to variables or values between clusters can be shown in the SQL server environment or they can be calculated with statistical method. In the last case we assume each cluster constitutes a Gaussian/ Multinomial distribution on a continuous/discrete attribute. The parameters of these distributions are then calculated for each cluster. For each specific attribute value, we can next evaluate its probability.

The easiest way to obtain classes (profiles) of learners is to apply supervised classification. This approach however needs to prepare a representative training set consisting of pairs of attributes vectors describing the student and the class he/she belongs to. This solution is described in [13], [14]. The presented there *Profile Rules Extractor* is based on a decision tree algorithm. Decision tree is a classifier in the form of a tree structure, where each node is either:

- a *leaf node* - indicates the class of examples, or
- a *decision node* - specifies a test to be carried out on a single attribute-value, with one branch and sub-tree for each possible outcome of the test.

1. For each attribute *a* of feature vectors find the normalized information gain from splitting on *a*
2. Find the attribute *a_best* with the highest normalized information gain
3. Create a decision *node* that splits on *a_best*
4. Recurse on the sublists obtained by splitting on *a_best*, and add those nodes as children of *node*

Fig. 4. The pseudocode of C4.5 algorithm of decision tree

A decision tree can be used to classify an example (the pattern that was not used in training) by starting at the root of the tree and moving through it until a leaf node. It gives classification of the pattern. The first popular decision tree algorithm *ID3* was proposed by Quinlann [25], [52]. Improvements of *ID3* was proposed by Quinlann and it is known as the *C4.5* algorithm. Its pseudocode is shown in Fig. 4. Some of these improvements are: handling both continuous and discrete values, handling training data with missing attribute values, handling attributes with differing costs. The latest version *C5.0* proposes trees pruning after creation. It is significantly faster than *C4.5* (several orders of magnitude), uses less memory, delivers smaller decision trees. The two most common measures for tests performed in decision nodes to find the best attribute to split (*a_best*) are entropy and the Gini index.

In the context of personalization the *C4.5* algorithm was also applied by Lee [27]. Tree production has three phases in their algorithm. In *Phase I*, an initial and large tree is created from the sets of examples according to attribute selection measures. In *Phase II*, an error-based pruning method is used to simplify the tree by discarding one or more sub-trees and replacing them with leaves or branches. In *Phase III*, the classification performance of the decision tree is tested by analyzing the number of correctly and incorrectly classified instances. The number of correct classified instances determines whether the decision tree can be applied to the data sets, or whether further preparation will be necessary.

The *Profile Extractor* [13], [14] is based on *C4.5* algorithm. Its input is represented by the XML file that contains the personal and interaction data of the user. The subset of the instances chosen to train the learning system has to be preclassified by a domain expert. On this basis the *Profile Rules Extractor*, infer classification rule set, which is used to predict the user behavior in a system for unclassified patterns. The premise part of a rule is a series of tests, just like the tests at nodes in the classification path of a decision tree, while the conclusion gives the class that applies to instances covered by that rule.

The *Expert Rule* – a data mining tool, applied by Sun and others [43] to classify learners, can be identified with the *C4.5* algorithm. For easier understanding of the detailed records of classifying all observed groups, the results of the *Expert Rule* are transformed into a visual illustration of rules classification. All groups are well classified in the final sets by the *Expert Rule Classifier*. They also implemented a novel hierarchical clustering in their system. This grouping method based on data mining techniques is shown to produce more effective groups.

Chen and others [9] discover potential student groups (profiles) that have similar characteristics and reaction to a particular pedagogical strategy. They applied decision tree (*C5.0*) algorithm and data cube technology.

In [44] in order to obtain the profile of learner the authors used model closely related to the *Naive Bayes* for classification in which all features are treated as conditionally independent. The *Naive Bayes* classifier combines this model with a decision rule. It is based on Bayesian theorem. It is particularly suited when the dimensionality of the inputs is high. The *Naive Bayes* classifiers can handle an arbitrary number of independent variables (features) continuous or nominal. In this classification the aim is to find for each vector $\boldsymbol{v}=[v_1,v_2,..v_n]$ appropriate class c_j . The most probable class c_j is the one which maximizes the a'posteriori probability $p(c_j|v_1,v_2,...v_k)$, i.e., the product of conditional class density function $p(v_1,v_2, ..., v_k|c_j)$ and a'priori probability $p(c_j)$. Ueno [49] uses also applied Bayesian approach for online outlier detection. Outliers represent learners' irregular learning processes.

1. Generate the initial population of individuals
2. Calculate the fitness value for each individual in that population
3. Repeat on this generation until stop condition is met: (time limit, sufficient fitness achieved, etc.)
    1. Select the best-fit individuals for reproduction
    2. Create new individuals by applying crossover and mutation operations
    3. Evaluate the individual fitness of new individuals

Fig. 5. The simple genetic algorithm scheme

Chang and others [7] applied the well known *k-Nearest Neighbor* (k-NN) classification method to classify students according to their learning style. For *k-NN* classification, an object (described by vector of features) can be represented as a point in an *n*-dimensional space. If two similar objects belong to the same class, their distance is shorter than that of other samples that belong to other class. In order to classify an unclassified object the distance between it and the samples has to be calculated. This causes that for a huge amount of data the algorithm can be inefficient. To improve this situation the authors proposed an application of *Genetic Algorithm* (*GA*) to decrease the number of features taken into account while a distance is calculated for *k -NN* algorithm.

*GA*s are adaptive metaheuristics used as an optimization tool. The algorithm in in pseudocode is shown in Fig. 5. The power of *GA*s comes from their simplicity and elegance as robust search algorithms as well as from their power to discover good solutions rapidly for difficult high-dimensional problem. To find the solution by GA it is necessary to encode it in a chromosome, to define fitness function and genetic operators. The evolution starts with generation of an initial population. Then the individuals are evaluated and some of them are chosen to create offspring population. The new population is obtained on the basis of parent individuals after an application of genetic operations. An effective GA representation and meaningful fitness evaluation are the keys of the success in GA applications.

To classify learners according to their learning styles the authors in [7] applied enhanced k-NN method. The learning behavioral features of 117 elementary school students were collected and then classified by the proposed mechanism. The experimental results indicated that the proposed classification mechanism can effectively classify and identify students' learning styles.

5 Recommending System for e-Learning

The task of recommending system in an e-learning system is to propose for a learner an individual learning path based on the tasks already done by the learner and based on tasks made by other similar learners. These similar learners are included in the profiles. To achieve this recommendation for each profile we have to find a typical sequence of actions performed by this group in order to suggest such sequence for a new learner. This is made by personalization mechanism (Fig.1.), which contains a method discovering typical behavior of the group. Recommending systems can be perceived as support decision system that suggests decisions i.e., learning path for individual learner. Taking into account the subject matter of this chapter we are rather interested in recommending systems which are knowledge based. Their core components are knowledge base and the inference mechanism.

5.1 Searching Learning Path

After defining profiles each of them should be characterized by typical behavior of the learners belonging to this group. Typically the algorithms of associations rules mining are used in this case. There are several algorithms available. *Apriori* or *FP-Tree* [19] are some of the most popular pattern discovering algorithms. Association rules describe interesting relations between variables (features) in data set. In other words, they show attribute value conditions that occur frequently together in a given data set. They provide information in the form of IF-THEN statement, expressed by eq. 3.

$$IF\, v_i\, and\, v_k\, and\, v_l\, THEN\, v_z\,, \tag{3}$$

which says that if v_i and v_k and v_l appear then appears also v_z ; In the case of e-learrnig systems v_i refers to an action or link.

An association rule is characterized by two coefficients. The first one is called the *support*. It is simply the number of objects that include all items in the antecedent and consequent parts of the rule. (The support is sometimes expressed as a percentage of the total number of records in the data set).

The second one is known as the *confidence* of the rule. Confidence is the ratio of the number of records that include all items in the consequent as well as the antecedent (namely, the support) to the number of records that include all items in the antecedent. In other words, the confidence of the rules indicates how strong the rules are, whereas the support of the rules indicates their coverage.

This approach is applied in [51] to find which web pages are usually visited together in a session. The association rules discovered for each cluster characterize the navigation patterns of specific user groups. The association rules can serve as the knowledge models to predict the next navigation requests for future similar users grouped in a cluster. Those rules matched with sufficient confidence are fired, and the predicted items are added into the recommendation list in a sorted manner according to their decreasing confidence values.

Similar approaches with association rules are described in [23], [36], [37], [46]. In the last one all the sequential patterns are split in sequences of only two components. These obtained sequences can be considered as a rule with only one antecedent and one consequent. The antecedent represents the page in which the recommendation is shown and the consequent is the link recommended to the student.

Also [53] presents the idea of association rules to train the recommender agent to build a model representing the web page access behavior or associations between on-line learning activities. The association rule mining technique is applied on such transactions to discover associations between actions, associations between URLs and associations between actions and URLs, as well as associations between sequences of actions and/or URLs. The authors of the system introduced higher weights to the rules that have as a consequent a URL or a set of URLs that are frequently towards the end of a session. The rules referring to associate actions, are only kept if they have as a consequent an action that terminated successfully. The agent recommends online learning activities based on a learner's access history.

Markellou and others [30] discover association rules, using the *Apriori* algorithm. They propose an ontology, the role of which is to determine which learning materials are more suitable to be recommended to the user.

The classical *Apriori* algorithm is inefficient when the number of data is high, that is many researches are performed in order to improve its efficiency. For instance Lazcorreta [26] and colleagues proposed a modified version of the *Apriori* algorithm to the log files of a web site to help the users to select the best user-tailored links.

Other interesting approach is shown in [20]. This paper presents a genetic-based learning path generation. As the learning path a curriculum sequence is assumed. The idea of curriculum sequencing is to generate an individualized course for each student by dynamically selecting the most optimal teaching operation (presentation, example, question, or problem) at any given moment. A serial number from 1 to n is assigned to each curriculum, where n is the total number of curriculums in the curriculum database for the learning path generation. Thus, the assigned serial number of each curriculum is combined directly with the serial number of the successive curriculum as strings to represent the generated learning path for the genetic algorithm. The whole individual, is represented by the chromosomes of all curriculum parameters.

In ELENA [11] project matching is performed on the basis of inference rules which infer whether a document is recommended or filtered to particular user or not. Each document here is annotated by the topics covered in this document. Topics can be covered by sets of documents. However, only one attribute is provided for annotating resources with topics, but additional properties for these annotations might be useful in other contexts, so more attributes are planned in the future for annotating.

The system described in [38] is an example of genetic approach to find rules which describe relationships between the student's usage of the different activities and modules provided by this e-learning system and the final marks obtained in the courses. Here an evolutionary algorithm for the induction of fuzzy rules is proposed. The rules are in the form given by eq. 4:

$$IF\ v_1\ is\ LL_1^3\ and\ v_3\ is\ LL_3^1\ THEN\ Class_2\ , \tag{4}$$

where LL_i^k is a linguistic label of k-th fuzzy set for i-th feature. In this approach genetic algorithm (GA) searches for fuzzy classification rules. Each rule is encoded in one chromosome.

An interesting approach to define a learning path is applied in [3]. It proposes a feedforward neural network. This network is trained in a supervised way, which needs to prepare data as a set of pairs *<input vector, output vector>*. To prepare it items of course and corresponding responses of learners are presented to the instructor, who diagnosed learner's learning problems. Items of course and corresponding responses are given to input of neural network while the evaluation of a teacher created the output of the network for the neural network training. After training it is used to suggest a learning path for a new learner.

5.2 Knowledge Based Recommending Systems

Most intelligent recommending systems are knowledge based. These systems collect knowledge and they are able to infer on its basis. Expert systems and case based reasoning systems are the examples of knowledge based systems considered in artificial intelligence. But also these e-learning systems that are built on an agent paradigm in some point of view are knowledge based because each agent realizes its tasks on the basis of knowledge it poses. In this chapter we will focus on this classical view of knowledge based systems.

UMPERS [2] is a rule-based expert system. The creation of final recommendation material list based on student's requirements and on every answer given by the student is achieved. Sometimes the final recommendation material list is not absolute because it depends more or less on some critical answers received from the student during the modeling. An Expert System is a set of rules that can be repeatedly applied to a collection of facts about the world. Rules that apply are fired, or executed.

Knowledge based approach is also applied in [24]. The *Learning Assistant* supports users in a learning process by selection an individual learning path referring to the scope and type of didactic materials. Knowledge Base is a set of rules useful during plans generation and modification. The rules allow for actualization of some information concerning learners and didactic materials. The rule have forms: *IF premise THEN conclusion* and *IF premise THEN action*. All rules are divided into three groups, taking into account the goal of using them. The first group of rules are *metadata modifiers* – they modify metadata of didactic materials, groups of learners and learners. The second one are *selectors* – they are used for plans generation using metadata concerning learners, didactic materials and connections between them. The third group of rules are *modifiers* – they use also results of verifying tests. The proposed rules can be tuned or changed step by step as experience is gained.

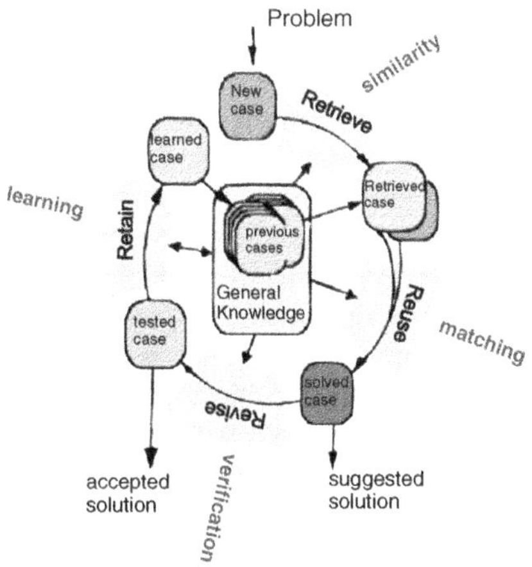

Fig. 6. Case Based Reasoning processing

Authors of the paper [6] use an extension of *Fuzzy Inductive Reasoning* methodology to extract comprehensible, actionable and reasonable set of rules describing the students' learning behavior. These rules can improve the learner behavior understanding and to provide valuable information to tutors about the course performance. *Fuzzy Inductive Reasoning* (FIR) uses fuzzy and pattern recognition techniques to infer system models and to predict their future behavior. It is well known that variations on fuzzy partitions have a direct effect on the performance of the fuzzy-rule-based systems. The FIR methodology is not an exception. The performance of the model identification and prediction processes of FIR is highly influenced by the discretization parameters of the system variables, i.e. the number of classes of each variable and the membership functions that define its semantics.

Huang [20] shows the *Case-Based Reasoning* (CBR) to develop a summarizing examination or assessment analysis. *CBR* is based on the intuition that new problems are often similar to previously encountered problems and, therefore, that past solutions may be used in the current situation. Typical processing in *CBR* consists of comparing a new problem (Fig.5.) which has to be solved with cases stored in the system in order to draw inferences about the problem and to guide decision making. CBR is an alternative for expert systems. Cases stored in the system are problems solved in the past. They create general knowledge about the problem to be solved. Knowledge in this case is much easier to acquire than for expert systems. In empirical study [20] the authors indicated that the proposed approach can generate appropriate course materials to learners based on individual learner requirements, and help them to learn more effectively in a Web-based environment.

5.3 Knowledge Representation in Recommending Systems

Knowledge is critical for intelligent systems such as those described in this chapter. The fundamental goal of a knowledge representation is to represent it in a manner as to facilitate reasoning (i.e, drawing conclusions) on its basis. Useful knowledge representation techniques in such systems are frames, rules and semantic networks. Semantic networks, sometimes called concept maps, represent knowledge in a form of graph of interconnected nodes. They include concepts, usually enclosed in circles or boxes of some type, and relationships between concepts or propositions, indicated by a connecting line between two concepts. Another characteristic of concept maps is that the concepts are represented in a hierarchical fashion with the most inclusive, most general concepts at the top of the map and the more specific, less general concepts arranged hierarchically below. Another important characteristic of concept maps is the inclusion of "cross-links."

Ontology specifies relevant concepts – the types of things and their properties – and the semantics relationships that exist between those concepts in a particular domain. They use a language with a mathematically well-defined syntax and semantics to describe such concepts, properties and relationships precisely. Ontology can be applied as a structured knowledge representation, capable of assisting the construction of a personalized learning path. The authors of [8] propose a novel genetic-based curriculum sequencing scheme based on a generated ontology-based concept map. This map can be automatically constructed by the pretest results of numerous learners to plan appropriate learning paths for individual learners. The experimental results indicated that the proposed approach could create high-quality learning paths for individual learners.

In ITES system knowledge is represented in the form of conceptual map [21]. Students learn new concepts and new connections between known already concepts. Some relations (connections) influence the ability of new concepts learning, they are called concept effect relationship. In ITES these relationships are used to show relationships between learning materials. Learning materials are visualized as diagram consisted of chapters, sections and subsections and key concepts that must be learnt. Defined connections between concepts and verifying possessed knowledge queries are guides for the further learning process. Mathematical formulas are proposed to point out a set of concepts that should be repeated in the learning process. The authors of the system claimed that experiments have shown that the system supports traditional education.

ELENA system [11], [12] supports e-learning in distributed learning environment using semantic network technology. It uses artificial intelligence techniques to provide friendly educational environment with a number of different functionalities. The aim of the system is to join personalization with open, dynamic network repository. The architecture of the system proposed by Dolog [12] is a basis for developing e-learning system working in distributed environment. Semantic nets can be used for metadata representation.

In [4] a rule based knowledge representation is proposed. Such representation is able to describe fundamental properties of didactic domain concepts. These rules are used to define relations among domain concepts and are applied in order to generate learning path. An ontology representation is used to store the admissible relations among the ontology domain concepts. These relations are as follows:

- HP (*Has Part*): HP(x, y_1, y_2, ..., y_n) means that the concept x is composed of the concepts y_1, y_2, ..., y_n ,
- R (*Requires*): R(x, y) means that to learn x it is necessary to have already learnt y.
- SO (*Suggested Order*): SO(x, y) means that it is preferable to learn x and y in this order and the following relation linking domain concepts and learning objects.

It is worth adding that a lot of information is hidden in learner's behavior, that can be represented by a set of operational and semantic attributes. Gu [17] describes user's activity ontology. It is possible to retrieve user's activity rules based on the previous user's activities. Each activity rule is represented in the form of conditional and qualified propositions, which are a specific kind of fuzzy propositions. Each fuzzy relation is represented by a membership value in [0, 1].

5.4 *Refining and Reorganizing the Learning Path*

Analyzing the sequences corresponding to the learning paths, the teacher can have an idea about what is the most general students' browsing behavior during their learning process.

A two-phase fuzzy mining and learning algorithm is described in [46]. It integrates an association rule mining algorithm – *Apriori*, with fuzzy set theory to find embedded information that could be fed back to teachers for refining or reorganizing the teaching materials and tests. In the second phase, it uses an inductive learning algorithm of the AQ family – AQR, to find the concept descriptions indicating the missing concepts during students' learning. The results of this phase could also be fed back to teachers for refining or reorganizing the learning path.

6 Intelligent Personalization – Critical Analysis

Table 1. contains a summary list of useful intelligent methods applied for personalization purpose in e-learning systems. We can mention here classification and clustering methods applied to the learner's data, to produce a learner's profile. Most clustering methods are based on a distance calculation so the proper definition of distance is essential for its successful application.

Table 1. Useful intelligent methods for personalization purpose in e-learning systems

method	task	stage of personalization	references
BP neural network	classification	learning path	[3]
k-NN	classification	profile creation -	[7]
decision trees	classification	profile creation -	[27], [43], [13], [9] [14]
Naive Bayes	classification	profile creation -	[49]
k-means	clustering	profile creation	[37], [55]
k-medoids	clustering	profile creation	[51]
SOM	clustering	profile creation	[24]
hierarchical clustering	clustering	profile creation	[47], [51], [43]
fuzzy logic	fuzzy clustering	profile creation – fuzzy rules – learning path	[48] [38], [46]
association rules	association	learning path	[23], [25], [30], [36], [37], [46], [53]
genetic algorithm	association	learning path, profile searching	[8], [20], [38]
knowledge based decision support system	recommending system	recommender	[24]
expert system	recommending system	recommender system	[2]
cased based reasoning	recommending system	recommender system	[20]

A disadvantage of simple *k*-means and *k*-medoids clustering algorithms is the need to define *k* before clustering. These methods are also sensitive for a high amount of data. An alternative for them can be SOM neural network, which offers also a visualization of clusters. There is no need to assign a number of clusters in advance in this case. It refers to hierarchical clustering as well. Genetic algorithm can be also profitable for clustering. However a proper tuning its parameters is not an easy task. Once they are found the technique gives good results.

We should notice that SVM classifiers are able to classify objects into two classes, only. Decision trees are sometimes more interpretable than other classifying method such as neural networks and SVMs because they combine simple questions about the data in an understandable way. So they can be easy transformed to the set of rules describing classification process and as such they describe classes in a comprehensible way. Unfortunately, small changes in input data can sometimes lead to large changes in the constructed tree. Decision trees are flexible enough to handle items with a mixture of real-valued and categorical features, as well as items with some missing features. They are expressive enough to model many partitions of the data that are not as easily achieved with classifiers that rely on a single decision boundary (such as logistic regression or support vector machines). However, even data that can be perfectly divided into classes by a hyperplane may require a large decision tree if only simple threshold tests are used. Decision trees naturally support classification tasks with more than two classes and can be modified to handle regression problems. Finally, once constructed, they classify new items quickly. Despite its simplicity, Naive Bayes can often outperform more sophisticated classification methods. An interesting approach could be an ensemble of classifiers or/and hybrid methods – some interesting proposals are presented in [9], [33].

To find a learning path suggested for a learner mostly association rules are applied. Most frequently the *Apriori* algorithm is implemented and most effort of authors is devoted to its modification. The discovered sequential patterns are used to create interesting recommendation for the students while they use the e-learning system. In our opinion the *FP-tree* [19] would be a better alternative. It uses a special representation of frequent patterns in a form of tree, which enables more efficient searching frequent sequential patterns having minimum support threshold. To find a learning path a genetic algorithm can be applied. It searches for rules that are frequent enough and have the assumed level of support coefficient. Searching a learning path is a difficult problem so in many cases they are given by an expert.

It is worth mentioning that knowledge included in recommending systems comes from experts or is acquired from data sets. Mostly it is expressed in the form of rules. This knowledge should be verified in order to find inconsistencies in the set of rules. Its maintenance is also a difficult task. This drawback does not exist in *Case Based Reasoning* systems (CBR), because knowledge in this case is represented by a set of cases (problems) for which we know the solution from the past. But here difficulties lie in defining similarity between cases and in principle when a new case should be added to the base. The number of cases in CBR decides about the response time so the choice between the number of cases stored in the system and the accuracy of solution found in the system is critical.

In e-learning systems usually there is a huge amount of data (e.g. logs) – and also these data sets can be used to improve effectiveness of e-learning process. However it is worth noticing that very rarely labeled data sets exist, which are required in supervised learning. This lead to the situation where the standard mentioned above techniques can give poor results. But the new learning paradigm can be useful in this case, so-called *Active Learning* (AL, or Learning By Questions). The main idea of AL relies in incremental learning using initially only a part of data which is annotated by a label of class and asking for label of data in the next steps only in problematic cases. More about AL idea, realization and applications can be found in [41], [10].

At the end of this critical analysis another important aspect of intelligent methods usefulness in e-learning systems personalization should be mention here. It is their scalability. Most systems that were described in this chapter are prototypes, which were not tested for large numbers of online users, e.g. for virtual campuses with thousands of students. From the analysis of their application to another fields it can be concluded that for instance scalability of *Apriori* algorithm is poor. Much better could offer *FP-tree* method. Also simple clustering methods such as k-*means* and k-*medoids* are inefficient for a huge data sets. So this aspect should attract more attention of researchers in future.

7 Conclusion

Intelligent techniques seems to be the good solution for automatic personalization in e-learning systems. Many intelligent methods exist that can be useful for this purpose. Which one to choose on a given stage is an issue of developer's experience in this field while tuning the parameters for each of them is crucial.

If we compare the actual survey with the older ones [38], [5] which contain overviews of data mining techniques a bit older (the last one till 2006), than it is clear that many new applications with the usage of intelligent techniques have arrived. This shows the power of intelligent methods which is perceived by e-learning system developers. Many researches have been performed in this area but much has to be done in order to built scalable systems based on these intelligent techniques.

References

[1] Agarwal, R., Deo, A., Das, S.: Intelligent agents in e-learning. SIGSOFT Softw. Eng. Notes 29(2), 1 (2004)

[2] Al Hamad, A., Yaacob, N., Al-Zoubi, A.Y.: Integrating 'Learning Style' Information into Personalized e-Learning System. IEEE Multidisciplinary Engineering Education Magazine 3(1), 2–6 (2008)

[3] Baylari, A., Montazer, G.A.: Design a personalized e-learning system based on item response theory and artificial neural network approach. Expert Systems with Applications 36, 8013–8021 (2009)

[4] Capuano, N., Gaeta, M., Micarelli, A., Sangineto, E.: An Intelligent Web Teacher System for Learning Personalization and Semantic Web Compatibility. In: Proceedings of the Eleventh International PEG Conference, St. Petersburg, Russia (2003)

[5] Castro, F., Vellido, A., Nebot, A., Mugica, F.: Applying Data Mining Techniques to e-Learning Problems. In: Lakhmi, C. (ed.) Evolution of Teaching and Learning Paradigms in Intelligent Environment, pp. 183–221. Springer, Heidelberg (2007)

[6] Castro, F., Vellido, A., Nebot, A., Minguillon, J.: Detecting a typical student behaviour on an e-learning system. In: I Simposio Nacional de Tecnologas de la Informacin y las Comunicaciones en la Educacin, Granada, pp. 153–160 (2005)

[7] Chang, Y., Kao, W., Chu, C., Chiu, C.: A learning style classification mechanism for e-learning. Computers & Education 53, 273–285 (2009)

[8] Chen, C., Peng, C., Shiue, J.: Ontology-based concept map for planning personalized learning path. In: IEEE Conference on Cybernetics and Intelligent Systems, pp. 1337–1342 (2008)

[9] Chen, G., Liu, C., Ou, K., Liu, B.: Discovering decision knowledge from web log portfolio for managing classroom processes by applying decision tree and data cube technology. Journal of Educational Computing Research 23(3), 305–332 (2000)

[10] Davy, M.: A Review of Active Learning and Co-Training in Text Classification, Computer Science Technical Report TCD-CS-2005-64, Trinity College Dublin (2005)
[11] Dolog, P., Nejdl, W.: Personalization in Elena: How to cope with personalization in distributed eLearning Networks. In: SINN 2003 eProceedings (2003)
[12] Dolog, P., Henze, N., Nejdl, W., Sintek, M.: Personalization in distributed e-learning environments. In: Proceedings of the 13th international World Wide Web conference on Alternate track papers & posters, pp. 170–179 (2004)
[13] Esposito, F., Licchelli, O., Semeraro, G.: Discovering Student Models in e-learning Systems. Journal of Universal Computer Science 10(1), 47–57 (2004)
[14] Esposito, F., Licchelli, O., Semeraro, G.: Extraction of User Profiles in e-learning Systems. In: Proceedings of I-KNOW'0, Graz, Austria, pp. 238–243 (2003)
[15] Fazlollahtabar, H., Mahdavia, I.: User/tutor optimal learning path in e-learning using comprehensive neuro-fuzzy approach. Educational Research Review, doi:10.1016/j.edurev.2009.02.001
[16] Gomes, P., Antunes, B., Rodrigues, L., Santos, A., Barbeira, J., Carvalho, R.: Using Ontologies for eLearning Personalization. Communication & Cognition 41(1&2) (January 2008)
[17] Gu, Q., Sumner, T.: Support Personalization in Distributed E-Learning Systems through Learner Modeling. In: 2nd Information and Communication Technologies, 2006. ICTTA 2006, vol. 1, pp. 610–615 (2006)
[18] Hamdi-Cherif, A.: Machine Elearning– Learning Agents and UML for Elearning Settings. International Journal of Education and Information Technologies 1(2), 51 (2008)
[19] Han, J., Kamber, M.: Data Mining: Concepts and Techniques, 2nd edn. Morgan Kaufmann, San Francisco (2006)
[20] Huang, M., Huang, H., Chen, M.: Constructing a personalized e-learning system based on genetic algorithm and case-based reasoning approach. Expert Systems with Applications 33, 551–564 (2007)
[21] Hwang, G.-J.: A conceptual map model for developing intelligent tutoring systems. Computer&Education 40(3), 217–235 (2003)
[22] Jing, C., Quan, L.: An adaptive personalized e-learning model. In: IEEE International Symposium on IT in Medicine and Education, 2008. ITME 2008, pp. 806–810 (2008)
[23] Khribi, M.K., Jemni, M., Nasraoui, O.: Automatic Recommendations for E-Learning Personalization Based on Web Usage Mining Techniques and Information Retrieval. In: Eighth IEEE International Conference on Advanced Learning Technologies, pp. 241–245 (2008)
[24] Kwasnicka, H., Szul, D., Markowska-Kaczmar, U., Myszkowski, P.: Learning Assistant- personalizing Learning Path in e-Learning Environments. In: 7 th International Conference Computer Information Systems and Industral Managament Applications (CISIM 2008), Ostrava, pp. 308–314 (2008)
[25] Kotsiantis, S.B.: Supervised Machine Learning: A Review of Classification Techniques. Informatica 31, 249–268 (2007)
[26] Lazcorreta, E., Botella, F., Fernaández-Caballero, A.: Towards personalized recommendation by two-step modified Apriori data mining algorithm. Expert Systems with Applications 35, 1422–1429 (2008)
[27] Lee, M., Chen, S., Chrysostomou, K., Liu, X.: Mining students' behavior in web-based learning programs. Expert Systems with Applications 36, 3459–3464 (2009)
[28] Li, J., Zaiäne, O.: Combining usage, content, and structure data to improve web site recommendation. In: International conference on ecommerce and web technologies, pp. 305–315 (2004)

[29] Lu, F., Li, X., Liu, Q., Yang, Z., Tan, G., He, T.: Research on Personalized E-Learning System Using Fuzzy Set Based Clustering Algorithm. In: Shi, Y., van Albada, G.D., Dongarra, J., Sloot, P.M.A. (eds.) ICCS 2007, Part III. LNCS, vol. 4489, pp. 587–590. Springer, Heidelberg (2007)
[30] Markellou, P., Mousourouli, I., Spiros, S., Tsakalidis, A.: Using semantic web mining technologies for personalized e-learning experiences. In: Proceedings of the web-based education, pp. 461–826 (2005)
[31] McAleese, R.: The knowledge arena as an extension to the concept map: Reflection in action. Interactive Learning Environments 6(3), 251–272 (1998)
[32] Mitchell, M.T.: Machine learning. McGraw-Hill, New York (1997)
[33] Myszkowski, P., Kwasnicka, H., Markowska-Kaczmar, U.: Data Mining Techniques in e-Learning CelGrid System. In: 7th International Conference Computer Information Systems and Industral Managament Applications (CISIM 2008), Ostrava, pp. 315–319 (2008)
[34] Quinlan, J.R.: Improved use of continuous attributes in C4.5. Journal of Artificial Intelligence Research 4, 77–90 (1996)
[35] Radenkovic, B., Despotovic, M., Bogdanovic, Z., Barac, B.: Creating Adaptive Environment for e-Learning Courses. JIOS 33(1), 179–189 (2009)
[36] Rigou, M., Sirmakessis, S., Tsakalidis, A.: Integrating Personalization in E-Learning Communities. Journal of Distance Education Technologies 2(3), 47–58 (2004)
[37] Romero, C., Porras, A., Ventura, S., Hervás, C., Zafra, A.: Using sequential pattern mining for links recommendation in adaptive hypermedia educational systems. In: International Conference Current Developments in Technology-Assisted Education, pp. 1015–1020 (2006)
[38] Romero, C., González, P., Ventura, S., del Jesus, M., Herrera, F.: Evolutionary algorithms for subgroup discovery in e-learning, A practical application using Moodle data. Expert Systems with Applications 36, 1632–1644 (2009)
[39] Romero, C., Ventura, S.: Educational data mining: A survey from 1995 to 2005. Expert Systems with Applications 33, 135–146 (2007)
[40] Romero, C., Ventura, S., Delgado, J.A., de Bra, P.: Personalized Links Recommendation Based on Data Mining in Adaptive Educational Hypermedia Systems. In: Duval, E., Klamma, R., Wolpers, M. (eds.) EC-TEL 2007. LNCS, vol. 4753, pp. 292–306. Springer, Heidelberg (2007)
[41] Ruiz, M., Díaz, M., Soler, F., Pérez, J.: Adaptation in current e-learning systems. Computer Standards & Interfaces 30, 62–70 (2008)
[42] Sun, C.-T., Chou, C.: Experiencing coral: design and implementation of distant cooperative learning. IEEE Transactions on Education 39(3), 357–366 (1996)
[43] Sun, P., Cheng, H., Lin, T., Wang, F.: A design to promote group learning in e-learning:Experiences from the field. Computers & Education 50, 661–677 (2008)
[44] Talavera, L., Gaudioso, E.: Mining student data to characterize similar behavior groups in unstructured collaboration spaces. In: Workshop on artificial intelligence in CSCL. 16th European conference on artificial intelligence, pp. 17–23 (2004)
[45] Tian, F., Zheng, Q., Gong, Z., Du, J., Li, R.: Personalized Learning Strategies in an intelligent e-Learning Environment. In: Proceedings of the 2007 11th International Conference on Computer Supported Cooperative Work in Design, pp. 973–978 (2007)
[46] Tsai, C.J., Tseng, S.S., Lin, C.Y.: A Two-Phase Fuzzy Mining and Learning Algorithm for Adaptive Learning Environment. In: Alexandrov, V.N., Dongarra, J., Juliano, B.A., Renner, R.S., Tan, C.J.K. (eds.) ICCS-ComputSci 2001. LNCS, vol. 2074, pp. 429–438. Springer, Heidelberg (2001)

[47] Tzouveli, P., Mylonas, P., Kollias, S.: An intelligent e-learning system based on learner profiling and learning resources adaptation. Computers & Education (2007), doi:10.1016/j.compedu.2007.05.005
[48] Tzouveli, P., Mylonas, P., Kollias, S.: SPERO - A Personalized Integrated E-Learning System. In: Conference proceedings of IADIS, pp. 34–22 (2004)
[49] Ueno, M.: Online outlier detection system for learning time data in e-learning and its evaluation. In: International conference on computers and advanced technology in education, pp. 248–253 (2004)
[50] Wallace, M., Akrivas, G., Stamou, G., Kollias, S.: Representation of user preferences and adaptation to context in multimedia content – based retrieval. In: Proceedings of SOFSEM, Milovy, Czech Republic (2002)
[51] Wang, F.-H., Shao, H.-M.: Effective personalized recommendation based on time-framed navigation clustering and association mining. Expert Systems with Applications 27, 365–377 (2004)
[52] Winston, P.H.: Artificial Intelligence, 3rd edn. Addison-Wesley, Reading (1992)
[53] Zaïane, O.R.: Building a Recommender Agent for e-Learning Systems, Computers in Education, 2002. In: Proceedings of International Conference on Computers in Education, vol. 1, pp. 55–59 (2002)
[54] Zeng, Q., Zhao, Z., Liang, Y.: Course ontology-based user's knowledge requirement acquisition from behaviors within e-learning systems. Computers & Education 53(3), 809–818 (2009)
[55] Zhang, K., Cui, L., Wang, H., Sui, Q.: Improvement of Matrix-based Clustering Method for Grouping Learners in E-Learning. In: Proceedings of the 11th International Conference on Computer Supported Cooperative Work in Design (2007)

Fuzzy ECA Rules for Pervasive Decision-Centric Personalised Mobile Learning

Philip Moore, Mike Jackson, and Bin Hu

Department of Computing, Birmingham City University, Birmingham, UK

Abstract. This chapter addresses personalisation in intelligent context-aware information systems. A personalized mobile learning system can be viewed as an information system related to the domain of education. Personalization requires the identification and selection of individuals; this can be achieved using an individual's profile (termed context). A context is inherently complex, its effective use representing a challenge that has to date not been adequately addressed. This chapter considers this challenge, identifies context-aware systems as intrinsically decision-centric, and introduces Fuzzy Event:Condition:Action (FEAC) rules as an effective approach to enable deterministic computational intelligence to be applied in intelligent context-aware personalised mobile learning systems. The FECA rules algorithm is presented with example implementations, an evaluation, and proof-of-concept. The chapter closes with a discussion, conclusions, open research questions, and consideration of future directions for future work.

1 Introduction

The principal motivation driving the work discussed in this chapter is personalization in decision-centric context-aware mobile information systems. There are many types of information system; a mobile learning system can be viewed as a particular type of information system applicable to the education domain. The goal of personalization is targeted service provision which offers the potential to address a number of important issues in mobile information systems including (1) *information overload*, (2) *personalized service provision*, (3) *inaccurate search results* (improvements are generally reflected in precision, recall and fall-out metrics), and (4) *interactive collaboration* in mobile systems (reflected in greater accuracy in the matching of users with shared interests).

Personalization of service provision requires the identification and selection of individuals based on their needs, desires, and interests (Moore *et al*, 2008). This requires the creation of an individuals profile (termed context) in human and computer readable form (Moore and Hu, 2007). Identification and selection implies acceptance (or alternatively rejection) of individual users based on their context, thus it is clear that the implementation of context reduces to a decision problem (Moore *et al*, 2008, 2009a, 2009b). A decision problem is characterized as one of selecting from a number of potential options (in this case individual users based on their context), the ability to arrive at a binary decision on the suitability of individual users as recipients for service provision can therefore be seen to be a pivotal requirement in context-aware pervasive systems.

F. Xhafa et al. (Eds.): Computational Intelligence for Tech. Enhanced Learning, SCI 273, pp. 25–58.
springerlink.com

Identification and selection of users [for personalized service provision] is fundamental in a context-aware mobile information system therefore all context-aware information systems are essentially decision-support systems. This analogy extends to ad-hoc wireless networks; such networks applying contextual information (albeit generally limited to location and identity) to dynamically configured network routing and the provision of services demanded by the users (Royer and Toh, 1999; Korkea-aho, 2000; Chen & Kotz, 2000).

Creating, updating and storing a context for use in a context-aware system is a dynamic and complex process, it has been shown (Moore and Hu, 2007; Moore *et al*, 2008) that the Semantic Web Technologies (W3C, 2009) implemented using Jena (in this research Jens2) (Jena, 2009) provides a basis for the effective creation of context definitions in a human readable and machine processable formalism. Context is however inherently complex, its effective use in intelligent context-aware systems representing a protracted and difficult challenge that has to date not been effectively addressed (Moore, 2009a, 2009b).

This chapter posits that intelligent context is required to effectively leverage the complexity of contextual information and introduces *Fuzzy Event:Condition:Action* (FECA) rules concept to enable the effective processing of contextual information with the ability to reach a binary decision in intelligent context-aware pervasive systems. The FECA rules build on and extend earlier work to develop a context framework with a context reasoning ontology in a prototype personalized mobile learning system (Moore and Hu, 2007; Moore *et al*, 2008). This work includes the development of the novel *Context Processing Rules* (CPR) (Moore *et al*, 2009b) which are predicated on a variation of the *Event:Condition:Action* (ECA) rules concept (Berndtsson & Hansson, 1995) extended using a quantitative approach in the context matching process (Moore *et al*, 2009b). The FECA rules extend the CPR using fuzzy extensions.

The remainder of this chapter is structured as follows: Personalization in information systems is discussed with technology enhanced learning, and pedagogic systems. A survey of stakeholder attitudes to context-aware mobile learning is presented with results and conclusions. Mobile systems are considered. Related research addressing context and context-awareness is considered with rules and, rule-based systems. The principal contribution of this chapter is the FECA rules concept. The FECA rules are presented with context-matching and fuzzy extensions. The FECA rules algorithm is presented with example implementations, an evaluation, and proof-of-context. The chapter concludes with a discussion, consideration of open research questions, and future work.

2 Personalization and Information Systems

An important issue in information systems is *information overload* caused by: (1) the exponential growth in data generation, (2) the increased ability of persistent storage systems to store the available data, and (3) developments in digital delivery technologies. Concomitant with these developments is the revolution in mobile technologies and systems with increasing ubiquity. Information overload is an issue shared by individuals, industry and academia alike and severely restricts the ability of users to be aware of, locate, access and utilize the mass of available data. Personalization offers the potential to play an important role in mitigating the worst effects of information overload.

Concomitant with the growth in the volume of data generated is the issue of data processing. Information systems are considered by Checkland & Holwell (Checkland & Holwell, 1998) and the traditional view of information as 'data processed resulting in information' is identified, this definition is however extended to include an intermediate stage termed "*capta*". *Capta* is defined as "data selected, created, or to which attention is paid" and involves the selection or creation of relevant data for processing into information useful (and of interest) to the user (Checkland & Holwell, 1998).

'Mobile and ubiquitous information access is discussed by Coppola *et al* (Coppola *et al*, 2003); information overload is considered and the concept of "relevance" introduced. Relevance refers to the degree to which the information is useful or of interest to potential users; therefore if a proportion of the information is 'relevant' potentially a greater proportion will be 'irrelevant'. The need to identify and identify and select appropriate recipients (users) for personalized (targeted) service provision is clear.

Information systems are a feature of a broad and diverse range of domains and systems; pedagogic systems being essentially information systems delivering academic content. The following section briefly considers the developments in tertiary education and the effects on technology enhanced learning in mobile systems.

2.1 Technology Enhanced Learning

There have been significant developments in tertiary education resulting in a growing interest in personalized educational provision; concomitant with these developments there has been a revolution in mobile technologies. The result of these developments has been a growing interest in personalized mobile learning.

There is a large body of research addressing academic and pedagogic issues, a prolific source for such work being the Association for Learning Technology (ALT, 2009). Academic and pedagogic research (generally) investigates socio-economic and demographic issues, course content, and the mode of delivery. Computational research into mobile systems and mobile learning typically (again generally) considers implementation issues related to mobility, mobile devices, service infrastructure, and service provision. While there are aspects common to research addressing academic / pedagogic and computational issues (see for example Goh & Kinshuk, 2004; Hokyoung & Parsons, 2008; Singh & Abu Bakar, 2007) a clear distinction can (generally) be drawn between the two research disciplines. A detailed exposition on the academic and pedagogic issues is beyond the scope of this chapter however examples of research addressing the topics include: Rushby (1998), MacKnight (1998), Cullen *et al* (2002), Gorrard *et al*, (2002), Mungania (2003), Long & Tricker (2004), Bull *et al* (2004), Liaw (2004), Dichev *et al* (2004), Komives (2005), and Scardamalia & Bereiter (2006) etc.

This chapter focuses on the computational issues as they apply to the use of context in Technology Enhanced Learning (TEL) to enable personalized service provision in intelligent context-ware pervasive mobile learning systems. We do not attempt an in depth discussion on the educational and pedagogic aspects in tertiary education, but rather hope to provide an overview of the issues with the motivation for personalized mobile learning. A survey-based evaluation of the attitudes of potential users of context-aware mobile learning systems is however provided with identification of the contextual information (context properties) that apply to a student in the domain of tertiary education.

2.2 Pedagogic Systems

There have been significant developments in tertiary education reflecting the evolving socio-economic and demographic make-up of the student population (Cullen *et al*, 2002; Gorrard *et al*, 2002); this has resulted in a growing interest in personalized educational provision – albeit currently at an embryonic stage (generally) in Virtual Learning Environments (VLE) such as 'moodle' (moodle, 2009).

'Moodle' has been successfully trialed and deployed at Birmingham City University, the initial trial being in the 2003-2004 academic year. Initially trialed in a dedicated module (A3E) within the Department of Computing 'Moodle' has been subsequently rolled out University wide over a distributed system in a wide area network (WAN) linking multiple campuses delivering a broad range of courses including computing, law, music, and the built environment. VLE's are considered in Britain & Liber (1999) and Cullen *et al*, (2002).

Birmingham City University is characterized by a very diverse student base in terms of attainment, demographics, cultural background, and needs from the HE experience; to reflect this a broad range of courses are offered from degrees to Masters and PhD level. The online presence has demonstrated positive benefits for students by providing learning *'on demand...anytime and anywhere'* as discussed in Goh & Kinshuk (2004), Rushby (1998), and MacKnight (1998).

There were, however, additional lessons related to staff involvement in a VLE. As discussed in Weller (2002) teaching staff need to feel comfortable with the increased demands of a VLE [manifested in the need to monitor modules delivered in a VLE and provide increased levels of interaction and feedback] for its use to be successful; this is not always the case which can be detrimental to overall process.

There has been significant research targeting mobile learning, for example, Goh & Kinshuk (2004) have considered issues in the implementation of learning modules in mobile learning using a simple case study and postulate that *"from e-learning to m-learning, mobile learning is going to be the next wave in the evolution of learning environments"*. Goh & Kinshuk (2004) in observing that a number of mobile learning and teaching applications have been deployed and evaluated (Luchini *et al*, 2002; Mifsud, 2002) conclude that while mobile learning can compliment e-learning by creating an additional channel of access for mobile users to engage in learning *"anytime and anywhere"* many issues regarding mobile learning have not been exhaustively researched.

Hokyoung & Parsons (2008) discuss recent developments in mobile learning techniques and related technologies, emerging innovations are considered with identification of appropriate pedagogies and technology usage methodologies. Singh & Abu Bakar (2007) describe the educational opportunities of teaching in a wireless classroom using notebooks and Personal Digital Assistants (PDA);

pedagogic issues are considered with the didactic aspects of mobile learning and the usability issues in accessing information on mobile devices.

Historically, computerized mobile learning systems have used networked and Internet-based approaches (generally termed 'e-learning'). Over time developments in mobile technologies have resulted in a blurring of the distinction between the e-learning and mobile learning (m-learning) concepts, the terms frequently being used interchangeably (Coppola *et al*, 2003; Kossen, 2005).

This ongoing trend adds significantly to the inherent complexity of context (Moore, 2009a) and its use in mobile learning systems. The application of contextual information (context properties) implemented in intelligent context-aware rule-based systems provides a realistic basis for the building of mobile learning systems. The FECA rules approach as discussed in this chapter provides an effective solution to the building of intelligent decision-centric context-aware mobile learning systems.

2.3 Stakeholder Attitudes to Context-Aware Mobile Learning

Personalized mobile learning involves many considerations that impinge on individuals including attitudinal and ethical considerations which can negatively affect the use and take-up of technology enhanced pedagogies. This aspect of personalization in context-aware personalized mobile learning was evaluated in a survey of stakeholders (staff and students) conducted at Universities in the UK and China.

The context factors (contextual information) applicable to a student in the domain of tertiary education (and used in the questionnaire) were identified in an interview process conducted with staff in the Department of Computing at Birmingham City University.

2.3.1 The Methodology

The context factors identified are presented in Table 1. The list is non-hierarchical being analogous to a menu from which applicable factors suitable to a domain-specific contextual design can be selected. The identification of the context factors was achieved using semi-structured interviews with staff drawn from the Department of Computing at Birmingham City University forming the sample population. Fifteen interviews were conducted, a breakdown shows that of the fifteen staff interviewed thirteen were active tutors, one was involved in student support, and one was employed in technical services. The tutorial staff interviewed included computing (technical) and business-oriented (soft systems) staff. A quantitative analysis shows that 66% were from a computing or technical background with 33% drawn from soft systems disciplines. Semi-structured interviews effectively addressed the data elicitation requirements and the associated issues identified, the principal issues being: (1) availability and timing constraints, (2) identification and access to the sample population, (3) ethical considerations.

The interviews were conducted by appointment on a one-to-one basis. Prior to the interview briefing documentation was provided to give the respondents an overview of the research, its aims, objectives and motivation. The time envisaged

Table 1. Contextual Information (context properties)

User Registered (Y /N)	Courses Taken
Student_ID	Engagement
Name	Progression
Date of Birth (D o B)	Cultural Background
Gender	Social Background
Address	Religion
Student Nationality	Ethnicity
Year (Enrolled Course)	Key Skills
Enrolled Course(s)	First Language
Registered Modules	Sexual Orientation
Mature Student (D o B)	Disabilities
Working (Y / N)	Web Surfing Patterns
Full Time	Connectivity
Part Time	Screen Size
Qualifications	Location (Where Contactable)
Experience (Academic)	Time Window (When Contactable)
Experience (Industrial)	Device Location (and therefore the user)
Projects	Device Availability (switched on / off)
Groups	Mode of Delivery (e.g., resource or message)
Interests	Resource Formatting (e.g., documents or files)

for each interview was 30 minutes, in actuality the time varied from approximately 30 to 75 minutes.

A potential issue identified at an early stage was the relatively small population size. The research design addressed this issue using the purposive sampling method (Robson, 2002). Purposive sampling calls for the use of populations with recognized knowledge and experience in the domain the research is designed to address, thus mitigating the potential for a biased result. The selected population satisfied the requirements of the purposive sampling method.

The factors were evaluated using a (38 question) questionnaire, the design being based on a variation of the Likert scale with four available options: (1) *strongly agree*, (2) *agree*, (3) *disagree*, and (4) *unsure*. An illustrative sample section of the questionnaire is presented in Figure 1. The questionnaire initially identifies the respondent as a student or tutor. The questions are set under four sections: (1) *personal*, (2) *academic*, (3) *mobile learning system*, and (4) *device*. The questions were framed using the identified context factors (contextual information). The questionnaire was subjected to a piloting process repeated using different (but similar) populations to that polled in the actual survey, the results and feedback being incorporated into the final questionnaire. The questionnaire was distributed to staff and students at Birmingham City University (UK) and Guilin University of Technology (P.R. China) to determine the degree to which the factors identified in the interview process are representative of other HE domains.

	Tutor	Student
Are you a tutor or a student	☐	☒

Context Factors	Strongly agree	Agree	Disagree	Unsure
Personal factors *Which of the following factors should be identified when defining a context*				
• Student record (user ID and the users name) – *required to identify each user uniquely*	☒	☐	☐	☐
• Address (student) – *relates to the distance from the university campus the student resides during his / her study periods e.g. term or semester*	☐	☒	☐	☐

student as an appropriate contact				
Academic factors *Which of the following factors should be identified when defining a context*				
• Engagement (student) – *relates to the level of involvement and active participation in the course / studies*	☐	☒	☐	☐
• Progression (student) – *relates to the academic progress as the course(s) and module(s) are completed*	☐	☒	☐	☐
• Learning style (student) – *relates to the students approach to study/learning e.g. a deep learner*	☒	☐	☐	☐
• Course enrolled (student) – *relates to the course on which a student is enrolled e.g. BSc / MSc etc*	☒	☐	☐	☐
• Course(s) taken (tutor) – *relates to the course on which a tutor teaches e.g. BSc / MSc etc*	☒	☐	☐	☐

Fig. 1. A sample section of an actual (4 page) questionnaire

2.3.2 The Analysis

A statistical analysis used aggregate response (N) values normalized to a maximum value of 1 with standard deviation (SD), correlation (C) and moving averages (MA). The results are cross-tabulated in Tables 2, 3 and 4; the results obtained from the statistical analysis were analyzed visually.

The formula for calculating the (N) values is: $\{N = (\sum R) / M\}$, where N = the aggregate response value, R = the sum total of the encoded responses, and M = the notional maximum value of the encoded responses. For example, given 10 'strongly agree' responses to a specific question the value of R is (10 * 4) which equates 40.

Applying the formula for **R** = 40 and also for a case where **R** = 30 the results are shown in (1) and (2) respectively:

$$\{ \mathbf{N} = (\sum \mathbf{R}) / \mathbf{M} \} \therefore \{ \mathbf{N} = 40 / 40 \} \therefore \{ \mathbf{N} = 1.00 \} \quad (1)$$

$$\{ \mathbf{N} = (\sum \mathbf{R}) / \mathbf{M} \} \therefore \{ \mathbf{N} = 30 / 40 \} \therefore \{ \mathbf{N} = 0.75 \} \quad (2)$$

It can be seen from the results obtained from applying the formula that in the first case the result is 1.00 and in the second case the result is 0.75. This demonstrates the overall level of agreement with a specific context factor.

2.3.3 The Results

The results obtained from the analysis of the questionnaire returns demonstrated a wide variation in response to each question, the SD (see Table 3) supporting this observation. The wide range of questionnaire responses are reflected in the data obtained in the interview process.

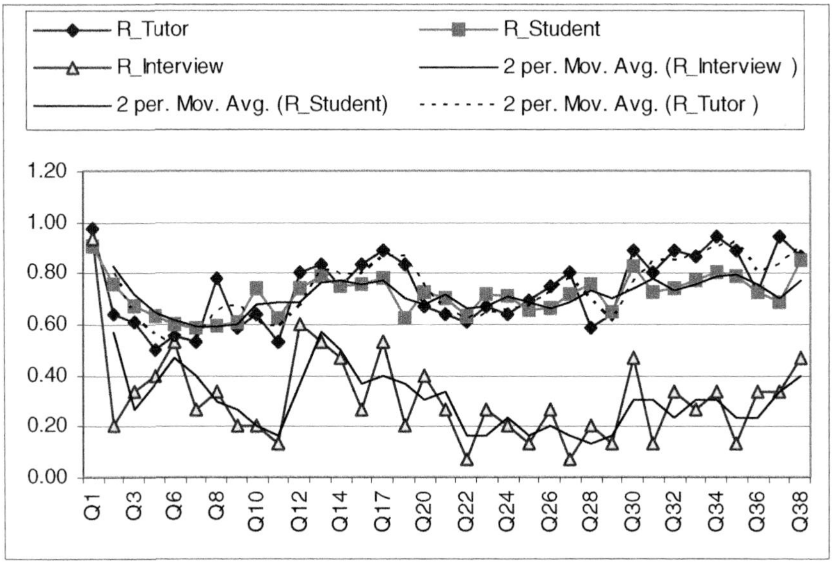

Fig. 2. Analysis of interview and questionnaire data

Table 2. Aggregate Response Values - (N) value ranges

	Max	Min	Avg
Tutor Responses	0.97	0.50	0.74
Student Responses	0.90	0.59	0.71
Interview Responses	0.93	0.07	0.31

Table 3. Standard Deviation - (SD) value ranges

	Max	Min	Avg
Tutor Responses	1.32	0.33	0.81
Student Responses	1.01	0.55	0.81

Table 4. Correlation Coefficient (C) values

Dataset Comparison	Coefficient
Interview data / Q_Tutor data	0.377069672
Student Responses	0.671195618
Interview Responses	0.480650532

To test for relationships between the interview, student, and tutor datasets and to quantify the strength of any relationship, correlation coefficients (C) were calculated. The (C) results were negative ranging from weak (interview/tutor data) through moderate (interview/student data) to strong (student/tutor data) (see Table 4). The results suggest that statistically significant relationships are present in the data.

The (N) values were computed for each question from the responses returned; a summary of the results are cross-tabulated in Table 2. The (N) values represent a measure of the overall level of agreement respondents expressed related to the use of each factor in a student context definition. The results fell into relatively narrow ranges for tutor and student responses, the interview results showing a similar maximum but widely differing minimum and average result. The wide differences identified are explained by the smaller sample used in the interviews as compared to the larger survey sample. With respect to the (MA), notwithstanding the widely differing (N) values identified, significant patterns and trends in all two datasets were observed. The relationships between the results obtained from the interview and questionnaire processes was analyzed graphically, the results are presented in Figure 2.

The results derived from the statistical and visual analysis support the observation that there is a level of statistical significance in the data. The analysis supports the conclusion that: (1) the wide variation in the identified factors noted in the interview process (and used in the development of the framework) is reflected in the questionnaire responses and (2) the factors identified in the interviews are representative of other larger populations.

3 Mobile Systems

Mobile systems are (generally) implemented in ad-hoc networks; such networks are (again generally) established and maintained dynamically by mobile and wearable devices (as discussed below) acting as network nodes with the nodes coming online and going off-line dynamically (Royer and Toh, 1999; Korkea-aho, 2000; Chen & Kotz, 2000). The result is dynamic change in network topology and membership; as such, ad-hoc networks are generally autonomous and self-configuring.

Ad-hoc networks can be viewed as inherently context-aware, contextual information being applied on a number of levels including: (1) establishing an ad-hoc network, (2) establishing and managing routing mechanisms, and (3) on an application level. Ad-hoc networks employ (in the main) spatio-temporal and identity based contextual information to establish dynamic connections between devices situated at specific logical or physical locations. A relatively common type of application is locating nearby resources including I/O devices such as printers and displays. Oxygen (MIT, 2009), a project to create an infrastructure of mobile and stationary devices connected to a self-configuring network is an example of ad-hoc network research.

Mobile computing devices are characterized by ubiquity and rapid evolution driven by dynamic developments in both hardware (computing devices) and software (operating systems and applications). Mobile devices (generally) fall into two broad classifications:

- Laptop and mobile computers: which whilst mobile are not generally useable whilst 'on-the-move'
- Wearable computing devices: which typically include mobile phones, smart phones and PDA's which can be used in a wider range of environments (albeit characterized by a diverse range of constraints) whilst 'on-the-move'

There has been significant convergence in mobile devices and their capabilities driven by developments in mobile technologies reflected in the size and capabilities of mobile computing devices, this has resulted in a blurring of the distinction between the two classifications. This is exemplified in the reductions in the development of the ultra mobile pc concept and the increased capabilities and power of wearable mobile devices with the related service infrastructure developments.

Concomitant with the developments in mobile systems is the increasingly distributed nature of information systems resulting in increased interactions between static networked systems and dynamic mobile systems. Context has been identified as inherently complex, its effective use in mobile systems presenting a significant increase in complexity as compared to static networked systems; this introduces the need for effective interaction between networked and mobile systems thus creating further complexity.

Given the broad range of contextual information that combines to form a context and the highly dynamic nature of the evolving mobile information systems environment the increased complexity must be handled effectively if true personalized service provision is to be realized. The current approaches to TEL however fail in that personalization in intelligent context-aware systems is not a feature; realization of this concept is a central requirement if true personalized learning in technology driven systems is to be achieved.

Developments in mobile computing have the potential to "free users from the desktop" (Abowd *et al*, 1997) however mobile devices have constrained capabilities when compared with desktop alternatives including: *restricted display*, *restricted bandwidth* and *fluctuating availability* (Lonsdale *et al*, 2003). Effective personalization in mobile information systems demands that these limitations are addressed.

Context offers the potential to mitigate the identified constraints and enable service provision compliant with user need in a format to suit the device type, capability and availability (Moore *et al*, 2009a. 2009b). The approach posited in this chapter uses a rule-based event driven system, which enables the realization of intelligent context processing in e-learning and m-learning systems whilst retaining the benefits and mitigating the negative aspects of such systems.

4 Context and Context-Awareness

Selection and identification is reliant on effective personalization, this requires that a users profile (termed context) be defined and described in computer readable form. The process of identification and selection requires that a context-aware system is capable of arriving at decisions based on a users context. Personalization therefore reduces essentially to a decision problem.

The first use of context in computer systems runs concurrently with the development of the ubiquitous computing paradigm as envisioned by Weiser (Weiser, 1991) and the emergence of mobile computing components in the early 1990's. Context forms an important element in the ubiquitous computing (alternative terms used have included 'pervasive computing' and 'ambient intelligence'); for example, in location-based services (such as context-aware tour guides) context is a pivotal function. This analogy clearly extends to personalized service provision in mobile information systems.

Context-awareness describes a concept in which an entity is defined by its 'context'; an entity is a person, place or physical or computational object. Context-awareness implements context to identify individuals to enable personalized service provision based on individual's 'situated role' which describes an individual and their prevailing environment. Context is however domain specific requiring the identification of the functions and properties (contextual information) specific to individual domains and applications. Context can be defined as: *'Contextual information (context properties) that combine to describe and characterize an entity and its situated role in human and computer readable form'.*

A very broad and diverse range of contextual information combine to form a context definition (Moore, 2009a, 2009b); in actuality, almost any information available at the time of an individuals interaction with a context-aware system can be viewed as contextual information including:

- The variable services demanded by users
- The diverse range of mobile devices and the associated service infrastructure(s)
- Resource availability (connectivity, battery condition, display, network, and bandwidth etc)
- Nearby resources (accessible devices and hosts including I/O devices)
- The physical situation (temperature, air quality, light, and noise level etc)
- The social situation (proximate information – who you are with, people nearby etc)

- Spatial information (location, orientation, speed and acceleration etc)
- Temporal information (time of the day, date, and season of the year)
- Physiological measurements (blood pressure, heart rate, respiration, and skin conductance etc)

In addition to the potential range of contextual information to be processed there is a need to accommodate two general types of context: a *static* context and a *dynamic* context (Moore *et al*, 2008):

- Static context (*customization*) relates to a use-case in which a users profile (context) is created manually, the user being actively involved in the process and having an element of control.
- Dynamic context (*personalization*) relates to a use-case in which the user is seen as passive, or at least somewhat less in control. In such a use-case the system monitors, analyses, and reacts dynamically to a users behavior and situated-role.

The two types of context are reflected in the two principal ways context is used, these are: (1) as a retrieval indicator (a static context) and (2) to tailor system behavior to match users system usage patterns (a dynamic context).

The characteristics of context identified when taken with the diverse range of mobile devices and service infrastructure clearly identify the inherent complexity of context.

5 Related Research

Research addressing context, context-aware applications can be traced from the early 1990's to the present day. Early research generally addressed office-based environments and include: *Active Badge* and *ParcTab* (Want *et al*, 1995), *In/Out Board* (Salber *et al*, 1999), and *Conference Assistant* (Dey and Abowd, 1999). Tourist information systems is a prolific field of research, examples include *Cyberguide* (Abowd *et al*, 1997), *Guide* (Davies *et al*, 1998), and *Crumpet* (Poslad *et al*, 2001). Other research projects have investigated *geographical applications* (Ryan *et al*, 1997), *memory aids* (Rhodes, 1997), *group interaction support* Ferscha *et al*, 2004), *motion capture* (Healey & Picard, 1998), *mobile learning* (Lonsdale *et al*,2003), *health monitoring applications* (Senses, 2003), *An Ambient Personalized Context-Aware Information system for Mobile Users* (Goker *et al*, 2004), and *knowledge management* (AKT, 2009). A comprehensive review of context-aware applications and systems can be found in Moore *et al* (2009).

Whilst the examples cited are not comprehensive the domains investigated point to: (1) the and the broad and diverse range domains investigated, (2) the potential complexity of context, and (3) the applications are generally information systems. There is however a more important conclusion; the documented research, in focusing (generally) on *spatio-temporal* and *identity* based information is limited in its scope, the rich and broad range of potentially available contextual information being under utilized and as such fails to address the inherent complexity that characterizes context in mobile information systems.

An important consideration for mobile information systems is the constrained nature of mobile devices. Addressing this issue requires the device-specific formatting of information which raises the issue of adaptation in the mode of delivery and presentation. Adaptation is discussed (Kinshuk & Goh, 2003) and the Multiple Representation approach is proposed to provide guidelines for content adaptation in e-learning and mobile learning environments. Adaptivity in mobile learning services is further considered in Kinshuk *et al* (2004). Adaptation in mobile information systems and accommodating the HCI and usability issues clearly adds significantly to the levels complexity in meeting the needs of users and accommodating the diverse range of mobile devices and the related service infrastructure.

There is a large body of research addressing rule-based systems in a broad and diverse range of domains (Hayes-Roth, 1985; Gonzales & Dankel, 1993; Mitchell, 1997) including database systems (Elmasari & Navathe, 1994; Berndtsson & Hansson, 1995), data mining and warehousing (Roiger *et al*, 1997), fuzzy logic (Zadeh, 1965; Klir & Yuan (1995; Berndtsson & Lings, 1995; Alvarado & Vázquez, 2004; Berkan & Truebatch, 1997), decision trees (Cerkvenik, 1997; Basak & Krishnapuram, 2005; Pop *et al*, 2005; Berger *et al*, 2006), artificial neural networks where classification of an input using threshold functions (Mitchell, 1997), and logical reasoning and inference (Jena, 2009; Horrocks *et* al, 2004).

Rules (generally) are conditional specifications that define instructions to: (1) instruct, (2) permit, (3) trigger, and (4) restrict or inhibit processes, functions, and actions. Rules when viewed from a computational perspective implement similar functionality using (generally) parameterized conditions in a machine (and ideally human) readable formalism. In considering rule-based systems there are four general types of rule, all of which (generally) test conditional relationships with a Boolean result:

- Derivation or Deduction rules: express conditional relationships (IF one set of statements is true THEN a second set of statements is also true), this rule structure generally relates to logical implication, a material conditional, or a Horn clause.
- Transformation rules: relate to truth (in for example one knowledge base) with truth in another. Transformation rules on (n-tuples) can be rewritten as derivation rules on (n+1-tuples) where the additional element specifies a knowledge base.
- Integrity Constraints: relate to truth where in inference "it must be true that…". General use relates to checking where "if it is not true…THEN (error)".
- Reaction or Event:Condition:Action (ECA) rules: relate to the notion of action (not just inference). For example, where an event occurs and the (IF) condition (the rule antecedent) is true the (THEN) action component (the rule consequent) of the rule is triggered.

The approach proposed in this chapter is predicated on the use of ECA rules with fuzzy extensions in an event driven rule-based system. In the FECA rules approach the condition(s) relate to contextual information (context properties – see Figures 1 and 2) and their values that define the body (antecedent) of the rule.

Testing the condition(s) (context matching) produces a binary result (the head or consequent of the rule) which triggers an action (either a binary result (a decision) or the firing of another rule). FECA rules employ an IF-THEN structure in the form: {<**On** <event> **IF** <condition> **THEN** <action>}.

The use of the IF-THEN rule structure is a feature of many rule-based systems; for example, when viewed from a decision support perspective IF-THEN structures have been shown to be effective in the generation of decision trees. The principle is exemplified in the illustration of the IF-THEN structure with conjunction (AND) logic operators to infer a conclusion:

$$\{\textbf{IF}\ (A1 \cap A2 \cap A3)\ \textbf{THEN}\ (\text{Class} \Rightarrow C1)\}$$

where {A1-A3} are logical statements which evaluate to a true or false (Boolean) result and {C1} is the class label enabling each class to be described by a set of rules and reasoned inference to be drawn (Basak & Krishnapuram, 2005). The adaptability of rules is discussed in Pop *et al* (2005) where the: *interpretability ... with decisions being representable in terms of a rule set"* is noted. This observation clearly points to the efficacy of rules in decision support systems and the reference to "*attribute value(s) or a combination of attributes...*" has a correlation with context properties and their values.

An interesting research project integrates the ECA rules concept with fuzzy ECA rule extensions to infer conclusions related to supply-chain management (Alvarado & Vázquez, 2004). The approach whilst interesting is limited in its scope, context representing a significantly more complex and challenging task than relatively simple supply-chain issues. The FECA rules approach is predicated on the W3C Semantic Web technologies (W3C, 2009); these technologies form the basis for the RDF Triggering Language (RDFTL) (Papamarkos *et al*, 2005). RDFTL applies the ECA rule concept and is a language for defining ECA rules on RDF repositories for centralized and distributed systems. RDFTL rules (generally) relate to *Derivation* or *Deduction* and *Reaction* (ECA) rule types. The research is however at an experimental stage and whilst providing an interesting application of the ECA rules concept the RDFTL approach along with similar approaches considered in Papamarkos *et al* (2005) fails to address the demands of context matching using the FECA rules.

The literature in demonstrating the adaptability of rules and rule-based systems supports number of conclusions:

1. IF – THEN rules have been shown to be an effective solution to arrive at binary decisions in decision-support systems however context matching (a fundamental concept in the FECA rules concept) represents a novel concept that is not addressed in the research.
2. Whilst ECA rule strategies are a generally well-understood (Gonzales & Dankel, 1993; Berkan & Truebatch, 1997, Mitchell, 1997; Pop et al, 2005 etc), fuzzy logic as it applies to fuzzy rule composition strategies is not (Berkan and Truebatch, 1997). The principal fuzzy rule composition strategies include: (1) Competitive rules (2) Cooperative rules (3) Weighted rules (4) Prioritized rules (5) Hierarchical rules and (6) Adaptive rules. A comprehensive discussion on the fuzzy rule composition strategies is provided in Berkan & Trubatch (1997)

however in summary given the inherent complexity of context no one rule strategy fulfils the demands of context matching, FECA rules incorporating strategic functions drawn from all the rule composition strategies; e.g., thresholds (Competitive rules), and weighting (Weighted rules).

3. Fuzzy logic as it applies to decision support addresses several classes of decision problem. A number of such classes are identified in Berkan & Truebatch (1997) however decision support in intelligent context-aware systems using fuzzy logic is an area underrepresented in the literature.
4. The use of rules with fuzzy extensions and the associated defuzzification based on a predetermined decision boundary (or threshold) (Berkan & Truebatch, 1997) provides the basis for context processing and the matching of input and output contexts (context matching – discussed in the following sections) in a decision-support system producing a binary decision relative to the suitability of an individual represented by the output context as measured against the input context.
5. While ECA rules (Berndtsson & Hansson, 1995) implemented using an IF-THEN structure have demonstrated the ability to provide an effective basis for general rule-based systems, the ECA rule concept alone fails to address the inherent complexity of context and the need to accommodate context processing and context matching including the partial matching issue which imposes similar issues to those encountered in decision support under uncertainty. Decision making under uncertainty is "possibly the most important category of decision problem" (Shackle, 1961; Klir & Yuan, 1995) and represents a fundamental issue in intelligent context-aware decision-support systems.
6. Fuzzy logic has demonstrated the capability to address cases where 'crisp sets do not apply (Klir & Yuan (1995; Berkan & Trubatch, 1997), context matching, which requires partial matching (as discussed in this chapter) being such a case. The ability of fuzzy logic when used as an extension to ECA rules provides the potential to achieve effective context matching.

In summary the literature has shown that an ECA rules strategy with fuzzy extensions has the potential to address uncertainty (and therefore context matching with partial matching). Notwithstanding the reservations expressed the research (Alvarado & Vázquez , 2004; Papamarkos *et al*, 2005) is interesting and confirms the potential of the ECA rules concept when used in conjunction with fuzzy extensions to address the inherent complexity and challenging nature of context-aware decision-support systems.

6 Context Processing

Context processing is predicated on the novel *Context Process Model* (CPM) (Moore *et al*, 2008); Figure 3 graphically models the process.

Context processing includes the context matching function. Context management with context processing is discussed in Moore & Hu, (2007) and Moore *et al* (2009b); a discussion on context matching is presented in Moore *et al* (2009b).

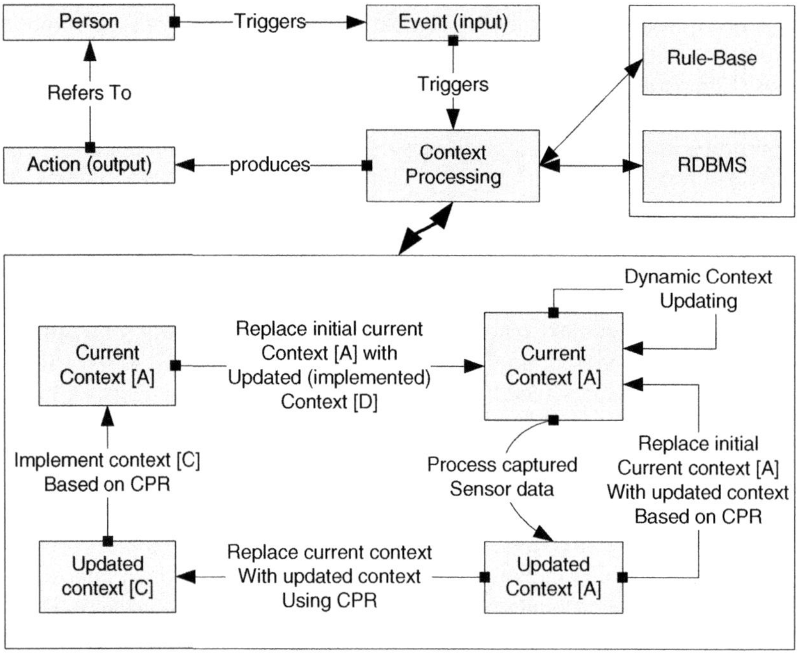

Fig. 3. The Context Process Model

The CPM implements the *Context Processing Rules* (CPR) and *Fuzzy Event*:*Condition*:*Action* (FECA) rules, the following section(s) address the creation and algorithmic approach to realize their implementation.

Prior to considering the CPR and FECA rules it will be useful to consider the nature and source of the contextual information that forms the input data. There are generally three data sources:

- Sensor derived data: are (generally) captured from location sensors such as RFID or Bluetooth sensing devices, such data are (generally) location and identity based.
- User supplied data: are (generally) captured from GUI sources and relate to user inputs which typically include resource input(s) (for distribution to suitably qualified individuals), collaboration query(s) (the aim being the identification of a suitably qualified advisor), or user input(s) of personal registration of preference data.
- System derived data: are (generally) the acquisition of spatio-temporal data (Moore, 2009a) and network data (such as login location). Such data is (generally) derived from servers and are important in enabling constraint satisfaction (user and system) and preference compliance and in the context-matching process.

7 The Fuzzy ECA Rules

This section presents the FECA rules with example implementations and an evaluation with proof of concept. A primary function of the FEAC rules is *context matching* (Moore *et al*, 2009b), a brief explanation and introduction of the context matching process is provided. The fuzzy extensions form an important element in the FECA rules algorithm, prior to presenting the algorithm fuzzy extensions are introduced. Following the evaluation there is a discussion on the posited approach with consideration of the open research questions.

7.1 Context Matching

A fundamental component in the context processing process is *context matching*. The *Context Matching* function is designed to access the *input* and *output* context definitions and using the context-matching algorithm determine if the *output* context (properties) are an acceptable match with the *input* context (properties). Figure 4 graphically models the partial matching issue the FECA rules algorithm is designed to address with Figure 5 graphically modeling the relationship between the input and output context properties.

Essentially, the context-matching process is one of reaching a binary decision as to the suitability of a specific individual based on his or her situated role (Moore *et al*, 2008, 2009b). Given that a perfect match is highly unlikely the context-matching algorithm must accommodate the partial matching issue along with a number of issues as discussed in Moore *et al*, 2009.

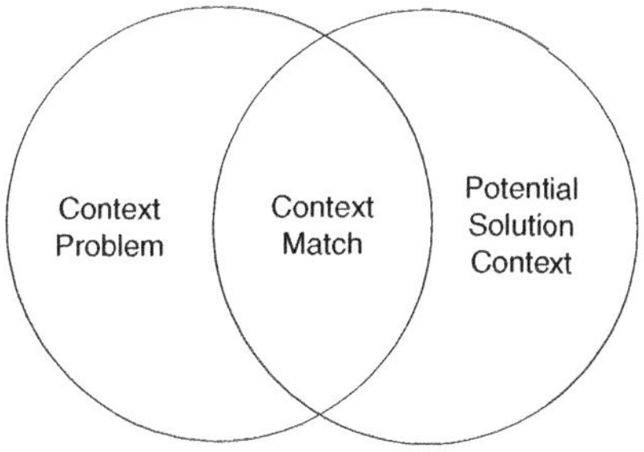

Fig. 4. The Context Matching Model and the Partial Matching Problem

7.2 FECA Rule Fuzzy Extensions

Conventional logic is generally characterized using notions based on a clear numerical bound (the crisp case); i.e., an element is (or is not) defined as a member of a set based on numerical parameters [1, 0]. A crisp set: relates to a condition in which an individual either belongs (or does not belong) to a set and set membership is in binary terms according to a "bivalent condition" (Zadeh, 1965). Fuzzy

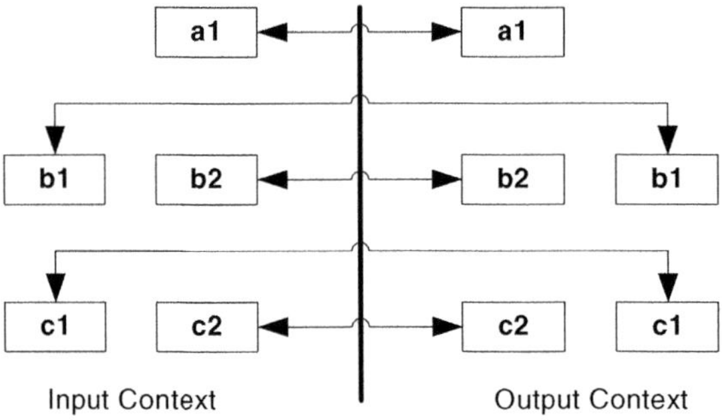

Fig. 5. The Context Matching (input / output) Model

set theory enables a variable measure of membership of a set defined using a membership function based on normalized values in the range [0…1]. An element that maps to a value of [0] indicates that the member is not included in a fuzzy set, an element mapping to a value of [1] indicates that an element is a fully included member of a fuzzy set. These mapping assumptions are central to the FECA rules concept as discussed in this paper. FECA rules are extensions of CPR (a discussion on CPR with an example implementation and an evaluation is presented in Moore *et al*, (2009b), the application of fuzzy extensions to the CPR is central in the FECA rules concept.

7.3 Fuzzy ECA (FECA) Rules

The FECA rules concept is an extension of the novel CPR. Space restricts a detailed discussion on the CPR concept and related topics including the approach to *rule assignment* however a detailed description can be found in (Moore *et al*, 2009b). In summary, CPR are predicated on the *Event:Condition:Action* (ECA) rules concept and apply the: {<**On** <event> **IF** <condition> **THEN** <action>} structure. CPR's use logic functions to implement the {**IF** <condition> **THEN** <action>} function using a binary structure similar to Horn style rules using: (1) a simple rule structure using a single context property (condition), and (2) a complex rule structure using multiple context properties (conditions) with conjunctions (AND), disjunctions (OR) and compliment (NOT) logic operators (Moore *et al*, 2009b). The CPR structure uses a quantitative approach with derived numerical values normalised in the range [0, 1] for each context property to enable the application of fuzzy extensions with defuzzification and enable binary decisions to be realized.

7.4 The FECA Rules Algorithm

Using the context properties (a1, b1, b2, c1, c2) shown in Figure 3 the context-matching algorithm is as follows where:

- **w** = [0.1…1.0] – the weight applied to a specific context property
- **e** = [0, 1] – the numerical (binary) evaluation for an individual context property match in which a true match = [1] and a false match = [0]

- **av** = $\prod$(**e** * **w**) – the *Actual Value* representing the degree to which each output context property matches the related input context property in the range [0.0…1.00]
- **sv** = $\sum$(**av** ($a_1 \ldots a_n$) – the sum of the *Actual Values* (**av**)
- **mv** = $\sum$($\mathbf{w}_1 \ldots \mathbf{w}_n$) – the *Maximum* (potential) *Value* for a perfect context match (the sum of the weights (**w**)
- **rv** = (**sv / mv**) – the resultant value, the degree to which the output context is a match for the input context in the range [0.0…1.0]
- **t** = the preset threshold value in the range [0.1…1.0]

7.4.1 The Algorithm

1. Evaluate the context match for each individual context property:

   IF (a_1(input)) ***equalTo*** (a_1(output)) **THEN e** = [1]

   IF (a_1(input)) ***notEqualTo*** (a_1(output)) **THEN e** = [0]
2. Obtain and set the weighting (**w**) for each context property:

   w = a_1[0.1…1.0]
3. Apply the weighting (**w**) to the value as derived from (2):

   IF e(a_1) = [1, 0] **THEN av** = (**e** * **w**)

   IF e(a_1) = [0] **THEN av** = (**e** * **w**)
4. Sum the values derived from the matching process:

   sv = $\sum$(**av** (a_1) + **av** (b_1) + **av** (b_2) + **av** (c_1) + **av** (c_2))
5. Compute the potential maximum value for the context-matching process:

   mv = $\sum$(**w**(a_1) + **w**(b_1) + **w**(b_2) + **w**(c_1) + **w**(c_2))
6. Compute the resultant value {**rv**} for testing against threshold value (**t**):

   rv = (**sv / mv**)
7. Using the preset threshold value (**t**) determine if the output (potential solution) context definition is a suitably qualified match with the input (problem) context:

   IF (**rv** >= **t**) **THEN** context-match = true [1]

   IF (**rv** < **t**) **THEN** context-match = false [0]

The algorithm set out above enables the realisation of a binary decision (i.e., an individual is classified as: (1) a suitably qualified recipient [1] or (2) not a suitably qualified recipient [0] to receive a resource or as a partner in a collaboration.

7.5 Implementation and Proof of Concept

This section presents two example implementations to provide an evaluation of the FECA rules concept and provide a proof of concept. The FECA rules algorithm is an extension of the *Context Processing Rules* (CPR) algorithm, a discussion on the CPR with an example implementation, results, and evaluation of the context matching process using the CPR in an interactive online collaborative interaction can found in Moore *et al* (2009).

Implementation One: Using the context properties shown in Figure 3 an implementation of the context-matching algorithm is as follows where the values for ($\mathbf{w}(a_1 \ldots c_2)$) are set out in (1), the values for ($\mathbf{e}(a_1 \ldots c_2)$) are set out in (2) and the threshold value ($\mathbf{t}$) is defined in (3):

$$\mathbf{w}(a_1) = 0.7,\ \mathbf{w}(b_1) = 0.8,\ \mathbf{w}(b_2) = 0.7,\ \mathbf{w}(c_1) = 1.0,\ \mathbf{w}(c_2) = 0.9 \quad (1)$$

$$\mathbf{e}(a_1) = [1],\ \mathbf{e}(b_1) = [0],\ \mathbf{e}(b_2) = [1],\ \mathbf{e}(c_1) = [1],\ \mathbf{e}(c_2) = [1] \quad (2)$$

$$\mathbf{t} = [0.65] \quad (3)$$

Applying the values set in (1), (2), and (3) results in a computation as follows:

$$\mathbf{av} = \Sigma(\ \mathbf{e}(a_1 \ldots a_n)\ *\ \mathbf{w}(a_1 \ldots a_n)\ )$$

$$\mathbf{sv} = \Sigma(\ \mathbf{av}(a_1) = ([1] \text{ x } 0.7) = [0.7]\ +\ \mathbf{av}(b_1) = ([0] \text{ x } 0.8) = [0.0]\ +$$
$$\mathbf{av}(b_2) = (\ [1] \text{ x } 0.7) = [0.7]\ +\ \mathbf{av}(c_1) = ([1] \text{ x } 1.0) = [1.0]\ +$$
$$\mathbf{av}(c_2) = (\ [1] \text{ x } 0.9) = [0.9]\ )$$

Based on the resultant values derived from applying the respective weightings to the Boolean results obtained in the evaluation of the equality test the normalised value for ($\mathbf{av}$) is:

$$\mathbf{sv} = (\mathbf{av}\ (a_1) = 0.7 + \mathbf{av}(b_1) = 0.0 + \mathbf{av}(b_2) = 0.7 + \mathbf{av}(c_1) = 1.0 + \mathbf{av}(c_2) = 0.9)$$

$$\mathbf{sv} = 3.3$$

To derive the maximum potential value $\{\mathbf{mv} = \Sigma \mathbf{w}_1 \ldots \mathbf{w}_n\}$

$$\mathbf{mv} = (\mathbf{w}(a_1) = 0.7 + \mathbf{w}(b_1) = 0.8 + \mathbf{w}(b_2) = 0.7 + \mathbf{w}(c_1) = 1.0 + \mathbf{w}(c_2) = 0.9)$$

$$\mathbf{mv} = 4.1$$

Given that ($\mathbf{mv} = 4.1$): ($\mathbf{rv} = \mathbf{sv} / \mathbf{mv}$) $\therefore$ ($\mathbf{rv} = 3.3 / 4.1$) $\therefore$ ($\mathbf{rv} = 0.80$).

The resultant value for ($\mathbf{rv}$) is 0.80 (to two decimal places) therefore as ($\mathbf{rv}$) is greater than or equal to the preset threshold value ($\mathbf{t}$), the result of the context match is ***true*** [1]. The individual **IS** therefore a suitably qualified individual.

Implementation Two: Again, using the context properties shown in Figure 3 an alternative implementation of the context-matching algorithm is as follows where values for ($\mathbf{w}(a_1 \ldots c_2)$) are set out in (4), the values for ($\mathbf{e}(a_1 \ldots c_2)$) are set out in (5) and the threshold value ($\mathbf{t}$) is defined in (6):

$$\mathbf{w}(a_1) = 0.7,\ \mathbf{w}(b_1) = 0.8,\ \mathbf{w}(b_2) = 0.7,\ \mathbf{w}(c_1) = 1.0,\ \mathbf{w}(c_2) = 0.9 \quad (4)$$

$$\mathbf{e}(a_1) = [1],\ \mathbf{e}(b_1) = [0],\ \mathbf{e}(b_2) = [1],\ \mathbf{e}(c_1) = [0],\ \mathbf{e}(c_2) = [1] \quad (5)$$

$$\mathbf{t} = [0.65] \quad (6)$$

Applying the values set in (4), (5), and (6) in a calculation as set out in implementation 1 results in (**av** = 2.3) and (**mv** = 4.1) (the maximum potential value (a perfect context match) derived from ($\sum \mathbf{w}_1 \ldots \mathbf{w}_n$). Applying these values results in:

$$(\mathbf{rv} = \mathbf{sv} / \mathbf{mv}) \therefore (\mathbf{rv} = 2.3 / 4.1) \therefore (\mathbf{rv} = 0.56)$$

The resultant value for (**rv**) is 0.56 (to two decimal places) therefore as (**rv**) is less than the preset threshold value (**t**), the result of the context match is ***false*** [0]. The individual is therefore **NOT** a suitably qualified individual.

An analysis of the results demonstrates that a number of important attributes incorporated into the FECA rules algorithm address some important issues and open research questions identified in Moore *et al* (2009). Additionally, a number of significant conclusions can be supported.

The evaluation of the context properties (step 1) and their respective values initially results in a Boolean result in the range (**e**) = [1, 0]. The truth or falsity [1, 0] is retained when the weight applicable to a context property (**w**) is applied (step 3) in the (**e** * **w**) context-matching step. For example, if the result of the initial evaluation process is [1] (as in implementation 1) there will always be a positive result. Conversely, for an evaluation result of [0] (as in implementation 2) the false result is retained.

8 A Dataset Evaluation

The preceding sections have discussed the FECA rule algorithm with (two) example implementations to demonstrate proof-of-concept. The following sections present an evaluation using an illustrative dataset to show how the data structure is implemented and demonstrate context matching. The *data structure* (with explanatory notes) is presented with illustrative examples of the context properties that combine to create a dataset. This is followed by an evaluation using an input set of context properties with context matching using two context(s).

However, prior to discussing the evaluation it is necessary to introduce a number of special cases where, for example, *transitive* relationships apply to the **equalTo** and **notEqualTo** equality tests (step 1 in the FECA rules algorithm). The following introductory section considers these special cases.

8.1 Introduction

The FECA rules approach is predicated on the Semantic Web technologies with a data structure encoded using RDF/S plus OWL in an ontology-based approach (Moore *et al*, 2008). Implementation is realised in a rule-based system as discussed in the *Fuzzy ECA (FECA) Rules* section.

The rules must be capable of implementing a number of 'states' including *transitive*, *cardinality*, and *enumeration*, etc. OWL DL in combination with RDF/S has been shown to enable logical inference and the capability to enable these 'states' with decideability (Moore *et al*, 2008, 2009b).

The **equalTo** and **notEqualTo** relationships (step 1 of the FECA rules algorithm) involve complex relationship tests where context properties including (but not restricted to – the identification of the actual context property relationships represents a domain specific design consideration) **date_Time** and **date_of_Birth** properties are evaluated in the context-matching process. This complexity is best-illustrated using examples of context properties which also illustrate potential issues in the mitigation of constraint satisfaction, preference compliance.

Consider initially the **date_Time** property. Suppose that a user restricts h/her contact to between (say) 09.00 and 17.00 hours, this introduces the transitive greater than $(\geq)$ and less than $(\leq)$ relationships. For example, if it is (at the time of the context-matching run) 19.00 or 08.15 hours then the test for equality in false. This relationship however disguises further complexity in the context processing function; it is not the intention to deny the user the resource or opportunity for collaboration therefore the system must decide when to contact the individual and which mode of contact (email, text message, or the increasingly popular social networking applications) to use to make contact.

Consider further the **date_Of_Birth** property. This property applies not only to the process of identifying specific individuals (e.g., a **name** property may not be unique but **date_Of_Birth** combined **name** almost is. Combining the **name**, **date_Of_Birth**, and **user_ID** properties will for all practical purposes provide a unique identity. Where the **date_Of_Birth** property is used in, for example, the testing for a mature student (often defined as a student over a specific age, e.g., 21) the **date_Of_Birth** property is again transitive $(\geq)$ or $(\leq)$. The **equalTo** test in this example is based on the **date_Of_Birth** and the **current_Date** where:

IF ((**date_Of_Birth** + 21) $\geq$ (**current_Date**)) **THEN** context-match = true [1]

The examples demonstrate: (1) the potential complexity to be accommodated, and (2) the domain specific nature of context-aware system design. Similar examples can be envisaged for complex relationships involving *cardinality* and *enumeration*. RDF/S with OWL DL provides the capability to define the *greater_Than* and *less_Than* relationships with domain and range restrictions (Antoniou & Van Harmelen, 2003). The *Context Processing Rules* (CPR) (Moore *et al*, 2009b) with the FEAC rules in a rule-based system enable the relative complexity of the **equalTo** and **notEqualTo** relationships to be effectively implemented. The following sections present a typical dataset with the evaluation.

8.2 The Data Structure

The data structure is built using the RDF triples plus an additional context property that defines a *weight* property to implement mitigation of constraint and preference violations and the potential for users who may benefit from service provision to be omitted from the distribution. The RDF triple is constructed from the following components which form the basis for the context reasoning ontology:

- The *Subject* <S> - represents the OWL *Class* in the ontology, e.g., the *Class:Person* holds the properties that describe and define an individual.

- The *Predicate* <**P**> - represents the *Property(s)* defined in ontology Classes, e.g., a typical property could be 'user_ID', 'name', 'date_of_Birth', 'role', 'current_Date', 'course_Title', etc.
- The *Object* <**O**> - represents the value an RDF triple defines. An *Object* <**O**> can take on one of two value types: (1) a *Literal Value* (e.g., "name"), and (2) a *URL or URI* – defined as *Uv* in the Jena2 (Jena, 2009) denormalized database schema.
- The *Weight* <**W**> component defines the relative importance of a property, the primary function being to provide an effective basis upon which constraint satisfaction and preference compliance can be achieved and violations mitigated. As discussed in the section addressing the FECA rules algorithm the <**W**> property falls into the range {0.1 … 1.0}.

The Data Structure as set out in Table 5 and shows representative examples of the dataset using the RFD triple data structure (<S> <P> <O>) plus the additional component <W>. The context properties in Table 5 are not comprehensive but are intended to demonstrate typical context properties used in the data structure and the RDF triple components. There follows explanatory notes to describe the context properties and their usage in the context processing and matching function. Note the convention that the Class name <S> is *capitalized* and the Predicate <P> uses a *lower case* designation. Space restricts the inclusion of the full dataset - for a discussion on the ontology and data structure see (Moore *et al*, 2009b).

- **Note 1:** The **user_ID**, **name**, and **date_of_Birth** properties are unique identifiers. They are used in the context processing to identify individuals (and resolve potential clashes), implement policy rights and permissions and to create a reference point for each individual context-matching run. In practice, the ***equalTo*** and ***notEqualTo*** tests are potentially transitive $(\leq)$ or $(\geq)$ as considered in the introduction to this section. The <**W**> property does not apply to the **user_ID** and **name** properties as these properties are not used in the context matching process.
- **Note 2:** The **role** property is used in the context processing to implement policy rights and permissions each individual. The role property will generally have a single value however it in not inconceivable that there may be multiple values; e.g., an individual may be both a tutor and a student or administrator.
- **Note 3:** The **registered** property identifies a user as a member of the system. This is a Boolean property in the range {yes / no}. While the **registered** property is used in context processing it is not used in context matching as only registered members have access to the system therefore the <**W**> parameter does is not apply as the **registered** property is not used in context matching.
- **Note 4:** In context processing **date_Time** properties during the resource entry phase have a logging function; however in context matching they are matched to spatio-temporal related constraints and preferences. It is important to note that while the majority of the context property values are stored in a persistent state in a RDMS (in our case the Apache Derby JavaDB implemented in the Sun Microsystems Netbeans) the **date** and **time** property values are not stored but are system generated and converted into the **date** (dd/mm/yyyy) and **time** (hh/mm) format. The format will be dictated by the system used in the

implementation of the context-matching process e.g., 'Java' powered systems may vary from (for example) LISP or Prolog systems.

- **Note 5:** The **course_Title, course_Level** and **module_Title** properties are used in context matching to identify if a potential recipient is suitably qualified recipient for a resource and if the resource will meet the needs and interests of a potential recipient. This property type will typically have multiple values in a list format. The course_Title and module_Title are self explanatory, the course_Level represents the level (e.g., undergraduate, Batchelor, Masters, and P.hD) of an individuals academic achievement.
- **Note 6:** The interest_Group property is used in context matching to identify if a resource will be of interest to potential recipients based on their defined interests, BDI and web surfing patterns (Moore *et al*, 2008). This property type will typically have multiple values in a hierarchical list format.
- **Note 7:** The intended function of the current_Location and location_Class properties is to identify a users location to and match the proposed service provision to that location compliant with pre-defined constraints and preferences. For example, the location_Class property may define a location such as the "Library" as a "restricted" location or a lecture theatre as a "smart" location. These is properties will only have a single value, the possibility for errors in locations must be recognized processes to disambiguate location conflicts must be implemented.
- **Note 8:** The **device_Type** and **device_State** properties are descriptors of specific devices, **device_Type** addresses the device type e.g., *wearable devices mobile devices*. The **device_State** is a descriptor of, for example, the screen size, resolution, available bandwidth, battery state, device availability (e.g., on or off), and financial issues (the type of account users operate and its current state). There will be a range of available **device_Type** however selection will be dependent on a range of parameters including availability and resource type etc.
- **Note 9:** The **data_Entry** and **data_Read** properties are policy properties and define the rights and permissions granted to specific individuals (generally based on their role) by the system administrator. These properties are applied in context processing but are not applicable in context matching.
- **Note 10:** The **file_Type**, **file_Size**, **file_Name**, and **file_Location** are significant context properties and are descriptors of the resource being entered onto the system. The **file_Name** property is not used in the context matching; it however forms part of the output to successfully identified recipients. The **file_Type**, **file_Size**, and **file_Location** properties are used in context matching to identify the suitable mode in which to deliver the resource and relate to the available or specified device(s) and their current 'state'. For example, potential delivery modes include the file or document, a text message or email, a contact mode using the developing social networking applications, or a URI / URL identifying the location of the resource is located. The **file_Type**, **file_Size**, **file_Name**, and **file_Location** are particularly interesting properties given the developing interest in expanded modes of communication which includes Web 2.0, the experimental Web 3.0 (Jain, 2006; Pan *et al*, 2008; Paul *et al*, 2008), and the ongoing interest in 'mashup' technologies as discussed in Anjomshoaa *et al* (2009).

Having considered the data structure and introduced special cases such as transitive relationships we can now present a scenario-based evaluation using a dataset predicted on the context properties as defined in Table 5. The following sections set out the scenario and present the evaluation with results. The conclusions drawn from the evaluation are considered in the concluding discussion, open research questions and future work.

8.3 The Scenario

To perform the evaluation it will be useful to consider an actual document that may be of interest to a number of interest groups. The document to be used is described below using the context properties defined in the data structure. Note, the authors are identified in the example and while not relevant in this evaluation may in certain cases be relevant. The interest group as envisaged by the user entering the resource (the document) onto the system has set the following parameters which define the context properties that define the input context.

- **file_Name** Document title: "Efficient RDF Storage and Retrieval in Jena2"
- **authors**: Kevin Wilkinson, Craig Sayers, Harumi Kuno, Dave Reynolds
- **file_Size**: 121 KB
- **file_Type**: PDF
- **file_Level**: 7
- **mature_Student**
- **file_Location**: http://www.bcu.ac.uk/PTM
- **course_Title**: "Computing", "MBA", "M-Learning"
- **course_Level**: 7
- **module_Title**: "Database Systems", "information Systems" "M-Learning", "E-Learning", "E-Commerce", " Semantic Web", "Business Studies"
- **interest_Group**: "Computer", "Database", "M-Learning", "E-Learning", "E-Commerce"

The context properties and the related values set out above form the input data capture. The values are somewhat arbitrary and are intended to be illustrative and descriptive of a typical resource. There will be additional system generated properties as discussed in the following section which addresses the evaluation.

8.4 The Evaluation

The evaluation will entail the matching of the input context properties and the related values for two potential user instances, CM#1 and CM#2. The context properties, the related input and output values are set out in Tables 6 and 7.

The context properties [in the evaluation – see Table 6] are deliberately restricted to aid the retention of clarity; full details of the evaluation including the CPR (Moore *et al*, 2009b) and FECA rules that implement many functions such as transitive relationships, reasoning, and inference is impractical in the space available. The results of the initial context matching (a measure of the ***truth*** of ***falsity*** of the context property matching) are set out Tables 6 and 7 with the application of the weighting set out in Table 8.

Table 5. RDF/S data structure

<S>	<P>	<O>	DataType	<W>	*Notes*
Person	user_ID	Lv	String	n/a	note 1
	name	Lv	String	n/a	
	date_of_Birth	Lv	Date_Time	#0.75	
Role	role	Lv	String	{0.1 … 1.0}	note 2
State	registered	Lv	Boolean	{0.1 … 1.0}	note 3
Date_Time	current_Date	Lv	Date_Time	{0.1 … 1.0}	note 4
	current_Time	Lv	Date_Time	{0.1 … 1.0}	
Course	course_Title	Lv	String	#0.50	note 5
	course_Level	Lv	Integer	#0.50	
Module	module_Title	Lv	String	#0.80	
Interest	Interest_Group	Lv	String	#0.90	note 6
Location	current_Location	Lv	String	{0.1 … 1.0}	note 7
	location_Class	Lv	String	{0.1 … 1.0}	
Device	device_Type	Lv	String	{0.1 … 1.0}	note 8
	device_State	Lv	String	{0.1 … 1.0}	
Policy	data_Entry	Lv	Boolean	{0.1 … 1.0}	note 9
	data_Read	Lv	Boolean	{0.1 … 1.0}	
Document	file_Type	Lv	String	{0.1 … 1.0}	note 10
	file_Name	Lv	String	n/a	
	file_Location	Uv	String	{0.1 … 1.0}	
	file_Size	Lv	Float	{0.1 … 1.0}	
	file_Level	Lv	Integer	{1 … 8}	

Table 6. Context property match for CM#1

<P>	**Input Values**	**Output Values**	**Result** (match)
course_Title	"**Computing**" "MBA" "M_Learning"	"**Computing**"	[1]
course_Level	**7**	**8**	[1]
module_Title	"**Data Structures**" "Information Systems" "M-Learning" "E-Learning" "E-Commerce" "Semantic Web"	"**Data Structures**" "Java Programming" "AI"	[1]
mature_Student	**Age > 21** (over 21)	**15/10/1985**	[1]
interest_Group	"Computer" "Database" "Education" "E-Commerce"	"Travel" "Music" "Technology"	[**0**]

Table 7. Context property match for CM#2

<P>	**Input Values**	**CM#2**	**Result** (match)
course_Title	"Computing"	"Education"	[**0**]
	"MBA"		
	"M_Learning"		
course_Level	7	6	[**0**]
module_Title	"Database Systems"	"**E-Learning**"	[1]
	"Information Systems"	"M-Learning"	
	"M-Learning"	"English"	
	"**E-Learning**"	"Mathematics"	
	"E-Commerce"		
	"Semantic Web"		
mature_Student	**age>21(over 21)**	**23/04/1976**	[1]
interest_Group	**"Computer"**	**"Computer"**	[1]
	"Database"	**"Education"**	
	"**Education**"	"Java"	
	"E-Commerce"		

Table 8. The application of the weights

	<P>	**<W>**	**CM#1** <V>	<R>	**CM#2** <V>	<R>
a1	course_Title	#0.50	[1]	0.50	[0]	0.00
b1	course_Level	#0.50	[1]	0.50	[0]	0.00
b2	module_Title	#0.80	[1]	0.80	[1]	0.80
c1	mature_Student	#0.75	[1]	0.75	[1]	0.75
c2	interest_Group	#0.90	[0]	0.00	[1]	0.90
$\sum a1+b1+b2+c1+c2$		3.45		2.55		2.45
	Overall Result		[True]	[True]	[False]	[True]

8.5 *The Results*

Prior to considering the results of the context match it will be necessary to set the decision boundary (the threshold); for the purposes of this evaluation it will be set at a normative value of 0.70. Tables 6 and 7 demonstrate the initial step in the context matching process (as step 1 of the FECA rules algorithm) with the output being either *true* [1] or *false* [0]; (the matched properties are highlighted in bold). Comparing the input values with the output values for each of the example datasets as set out in the FECA rule algorithm and implementations one and two (in the *Implementation and Proof of Concept* section) produces the following results.

- For CM#1 the initial overall context match (prior to the application of the weights is [0.8], the initial result is therefore true [1].
- For CM#2 the initial overall context match (prior to the application of the weights is [0.6], the initial result is therefore false [0].

The above results reflect the application of the *Context Processing Rules* (CPR) (Moore *et al*, 2009b). Applying the weights as defined in Table 5 in the FECA rules algorithm however changes the result as shown in Table 8 where <V> = value of the initial context match, <W> = weight, and <R> = the result of applying the <W> value [<V> × <W>).

8.6 The Analysis

The results presented in Table 8 validate the conclusions identified in the *Implementation and Proof of Concept* section and support the intuition that the application of context property weighting addresses:

- The potential for users who would potentially benefit from the resource to be excluded from the distribution. This is demonstrated by the correction of the [false] result in the case of CM#2 to a [true] result to reflect h/her interests.
- The retention of a [true] or [false] result where such a result is obtained from the initial context matching. This result is a product of the prioritizing bias introduced by the application of weighting to the context properties to reflect their relative importance.

The context properties, the values used, and the weighting applied are arbitrary and are designed to demonstrate context processing and the matching of context properties. The parameters used are however based on a realistic mobile learning scenario in the domain of tertiary education. As previously noted, the actual setting of the weights to be applied is domain specific requiring domain specific research to identify: (1) the context properties that are relevant to a specific domain, and (2) the relative importance to be attached to each context property in a context property hierarchy.

As discussed in the explanatory notes in this section, constraint satisfaction and preference compliance are important considerations. The provision of hierarchical levels of importance conferred to context properties that define and describe constraints and preferences using the weighting approach provides an effective basis upon which constraint satisfaction and preference compliance can be realized.

9 Research Questions and Future Work

This chapter has considered personalisation in intelligent context-aware information systems and identified personalized mobile learning as a specific type of information system in the domain of tertiary education. Personalization requires the identification and selection of individuals; this can be achieved using an individual's profile (termed context). Context is however inherently complex, its effective use representing a challenge that has to date not been adequately addressed. This challenge has been considered, context-aware systems have been identified as intrinsically decision-centric, and FECA rules have been posited as an effective approach to enable deterministic computational intelligence to be applied in intelligent context-aware mobile learning systems.

The evaluations presented have supported the intuition that motivates this research that the posited approach provides an effective basis upon which intelligent context-aware systems can be built. The research to date and the development of the FECA rules which, while addressing some fundamental issues has however identified some further issues and open research questions, the principal issues being:

1. The '*weighting*' applied to context properties in the FECA rules algorithm while enabling effective context matching requires extending to accommodate 'real-world' context-aware systems. This issue relates to the ability to set the weights for specific context properties to accommodate users evolving preferences and constraints. The challenge is to enable the weights to be set initially using a default value and modified dynamically 'on-the fly' to reflect the evolving preferences and constraints.
2. A recognized issue in our research (and in context-aware systems research generally) is the potential for *anomalies* and *ambiguities* in the locating of individuals. For example, an individual's mobile device may register its location with data captured from a single sensor or multiple sensors; additionally, to a context-aware system a mobile device registers its owner in a specific location. Based on this assumption the owner of the device may not actually be with the mobile device thus creating an anomaly which requires disambiguation.
3. Context-aware systems are characterized by highly dynamic context factors such as complex preferences derived from web surfing activities. Developing an approach to address this issue remains an open research question.

To address these challenges investigations have identified statistical approaches implementing *prior* and *posterior* probabilities. Approaches considered include Bayesian methods; the *Naïve Bayes Classifier* (Mitchell, 1997) has been investigated as a potentially productive solution. Development of developing the optimal approach (including the possible combining of approaches) however remains an open research question.

1. Two example implementations have been presented with an additional evaluation using an actual illustrative dataset; testing of the proposed approach is now required to investigate the *weighting* and *threshold* levels. Investigations have failed to identify suitable benchmarks. Initially thresholds in the range 0.6 to 0.8 will be used, this is predicated on the results obtained in information retrieval research, an example of such research with an evaluation and results can be found in (Lang, 1995).
2. *Data description*: relates to the representation of user input to enable effective context matching. Approaches under consideration include: *Keywords* and *Natural Language Processing* (NLP). Initially, the selected option is *Keywords*, there are however issues relating to the keyword corpus, investigations having failed to locate a suitable corpus. The development of a suitable corpus of

keywords remains an open research question. Data entry using natural language is a desirable aim, *NLP* is however very challenging, computationally expensive, and remains an open research question.

3. *Database Schema*: the data storage must accommodate the richness of RDF and OWL DL, a data model typically found in online applications such as e-commerce systems which are predicated on datasets which are highly dynamic, constantly evolving, sparsely populated, and may contain numerous null values (Agrawal *et al*, 2001). Mobile information systems may conform to this paradigm. Contextual data in personalized mobile learning systems is however characterized by a dataset where data is (generally) dense, evolves slowly, and has few, if any, null values. Developing a database schema to accommodate these potentially conflicting requirements is a challenging general open research question as discussed in Agrawal *et al* (2001) and Wilkinson *et al* (2003). The initial implementations of the FECA rule approach will use a sub-optimal solution using the Jena2 'denormalized' relational database data schema.

This chapter has addressed personalisation in intelligent context-aware information systems and identified personalized mobile learning as a specific type of an information system in the domain of tertiary education. The issues that motivate personalization of learning in tertiary education have been introduced. The nature of context and its inherent complexity has been considered and the FECA rules concept has been posited as an effective approach to enable deterministic computational intelligence to be applied in intelligent pervasive context-aware personalised mobile learning systems.

References

Abowd, G.D., Atkeson, C.G., Hong, J., Long, S., Kooper, R., Pinkerton, M.: Cyberguide: A Mobile Context-Aware Tour Guide. Wireless Networks (3), 421–433 (1997)

Agrawal, R., Somani, A., Xu, Y.: Storage and Querying of E-Commerce Data. In: Proc. of VLDB (2001), `http://wotan.liu.edu/docis/lib/`

AKT, The AKT project (2009), `http://www.aktors.org/akt/`

ALT, ALT - the Association for Learning Technology (2009), `http://www.alt.ac.uk/index.html`

Alvarado, M., Vázquez, M.: Decision making automation based on fuzzy event-condition-action rules (2004), `http://delta.cs.cinvestav.mx/~matias/PDF/`

Antoniou, G., Van Harmelen, F.: Web Ontology Language: OWL. In: Staab, S., Struder, R. (eds.) Handbook on Ontologies. Springer, Germany (2003)

Anjomshoaa, A., Bader, G., Min Tjoa, A.: Exploiting Mashup Architecture in Business Use Cases. In: Procs. of the 2009 International Conference on Network-Based Information Systems (NbiS 2009), UIPUI, Indianapolis, USA, August 19-21 (2009)

Berkan, R.C., Trubatch, S.L.: Fuzzy Systems Design principles: Building Fuzzy IF-THEN Rule Bases. IEEE Press, Piscataway (1997)

Berndtsson, M., Hansson, J.: Issues in active real-time databases. In: Proc. of the 1st International Workshop on Active and Real-Time Database Systems (ARTDB 1995). Skovde, Sweden, Workshops in Computing, June 1995, pp. 142–157. Springer, London (1995)

Britain, S., Liber, O.: Pedagogical evaluation of virtual Learning environments (Tech Rep: 41) JISC Technology Applications Programme (October 1999)

Basak, J., Krishnapuram, R.: Interpretable hierarchical clustering by constructing an unsupervised decision tree. IEEE Transactions on Knowledge and Data Engineering 17(1), 121–132 (2005)

Bull, S., Cui, Y., McEnvoy, A.T., Wei Yang, R.E.: Roles for learner models. In: Proc. of the 2nd IEEE International Workshop on Wireless and Mobile Technologies in Education (WMTE 2004), Taoyuan, Taiwan, March 2004, pp. 124–128 (2004)

Checkland, P., Holwell, F.: Information, systems and information systems: making sense of the field. John Wiley & Sons, Chichester (2008)

Chen, G., Kotz, D.: Solar: an open platform for context-aware mobile applications (Tech. Rep.). Department of Computer Science, Dartmouth College, Hanover, NH, USA 03755. Dipartimento di Scienze dell'Informazione, Università of Bologna, Italy (2002)

Coppola, P., Mea, V.D., Gaspero, L.D., Mizzaro, S.: The concept of relevance in mobile and ubiquitous access. In: Crestani, F., Dunlop, M.D., Mizzaro, S. (eds.) Mobile HCI International Workshop 2003. LNCS, vol. 2954, pp. 1–10. Springer, Heidelberg (2004)

Cullen, J., Hadjivassiliou, K., Hamilton, E., Sommerlad, E., Stern, E. Report submitted to the Economic and Social Research Institute, London. Review of current pedagogic research and practice in the fields of post-compulsory education and lifelong learning. The Tavistock Institute (2002)

Davies, K., Mitchell, K., Cheverest, K., Blair, G.: Developing a context sensitive tourist guide (GIST Tech. Rep. G98-1). In: First Workshop on Human Computer Interaction with Mobile Devices

Dey, A.K., Abowd, G.D.: Towards a better understanding of context and context-awareness (GVU Tech Report GIT-GVU-99-22). Georgia Institute of Technology, Georgia, USA (1999)

Dichev, C., Dicheva, D., Arroyo, L.: Using topic maps for web-based education. Advanced Technology for Learning 1(1), 1–7 (2004)

Elmasari, R., Navathi, S.B.: Fundamentals of Database design, 2nd edn. Benjamin/Cummins, Redwood City (1994)

Ferscha, A., Clemens, H., Oppl, S.: Context awareness for group interaction support. In: Proc. of MobiWac 2004, Philadelphia, PA, USA, October 1, pp. 88–97 (2004)

Kinshuk, G.T.: Getting Ready For Mobile Learning. In: Cantoni, L., McLoughlin, C. (eds.) Proceedings of ED-MEDIA 2004 - World Conference on Educational Multimedia, Hypermedia & Telecommunications, Lugano, Switzerland, June 21-26, pp. 56–63 (2004)

Goker, A., Watt, S., Myrhaug, H.L., Whitehead, N., Yakici, M., Bierig, R.: An ambient personalised and context-sensitive information system for mobile users. In: Proc. of EUSAI 2004, Eindhoven, Netherlands, November 8-10 (2004)

Gonzales, A.J., Dankel, D.D.: The engineering of knowledge-based systems theory and practice. Prentice-Hall, NJ (1993)

Gorrard, S., Selwyn, N., Madden, L., Furlong, J.: Technology and lifelong learning: are we cutting IT? In: All-Wales Education Research Conference, Gregynog, July 3-5, 1993 (2002)

Hayes-Roth, F.: Rule-Based Systems. Communications of the ACM 28(9), 921–932 (1985)

Healey, J., Picard, R.W.: Startlecam: a cybernetic wearable camera. In: Proc. of the 2nd. International Symposium on Wearable Computers, Pittsburgh, Pennsylvania, October 19-20, pp. 42–49 (1998)

Hokyoung, R., Parsons, D.: Innovative mobile learning: techniques and technologies. Idea Group Inc. (IGI), Hershey (2008)

Horrocks, I., Patel-Schneider, P.F., Boley, H., Tabet, S., Grosof, B., Dean, M.: SWRL: A semantic web rule language combining OWL and RuleML (2004), http://www.w3.org/Submission/SWRL/

Jain, R.: Internet 3.0: Ten Problems with Current Internet Architecture and Solutions for the Next Generation. In: Proc. of the IEEE Military Communications Conference (Mil-com 2006), Washington DC, October 23-25 (2006)

Jena, Jena – A Semantic Web Framework for Java (2009), http://jena.sourceforge.net/index.html

Kinshuk, G.T.: Mobile adaptation with multiple representation as educational pedagogy. In: Uhr, W., Esswein, W., Schoop, E. (eds.) Wirtschaftsinformatik 2003-Medien-Markte-Mobilitat, pp. 747–763 (2003)

Kinshuk, K.P., Sampson, D.G.: Adaptivity in web and mobile learning ser-vices. Int. J. Cont. Engineering Education and Lifelong Learning 14(4-5), 313–317 (2004)

Klir, G.K., Yuan, B.: Fuzzy Sets Ana Fuzzy Logic Theory and Applications. Prentice Hall, London (1995)

Komives, S.R.: The Changing Nature of Work in Higher Education (2005), http://www.acpa.nche.edu/seniors/trends6.htm

Korkea-aho, M.: Context-aware applications survey (2000), http://users.tkk.fi/mkoreaa/context-aware.html#

Kossen, J.S.: Mobile e-learning When e-learning becomes m-learning (2005), http://www.palmpowerenterprise.com/issuesprint/issue200106/elearning.html

Lang, K.: NewsWeeder: Learning to Filter Netnews. In: Proc. of the 12th International Machine Learning Conference (ML 1995), Lake Tahoe, USA, pp. 331–339 (1995)

Shu-Sheng, L.: Considerations for developing constructivist web-based learning. International Journal of Instructional Media 31(3), 309 (2004)

Long, P., Tricker, T.: Do first year graduates find what they expect? In: British Educational Research Association Conference 2004, University of Manchester, September 16-18 (2004)

Lonsdale, P., Barber, C., Sharples, M., Arvanitis, T.N.: A Context Awareness Architecture for Facilitating Mobile Learning. In: Proc. of MLEARN 2003, London, UK, May 19-20, pp. 79–85 (2003)

Luchini, K., Curtis, M., Quintana, C., Soloway, E.: Supporting learning in context: extending learner-centered design to the development of handheld educational software. In: Milrad, M., Hoppe, H.U., Kinshuk (eds.) IEEE International Workshop on Wireless and Mobile Technologies in Education 2002, pp. 107–111 (2002)

MacKnight, C.B.: Electronic learning materials: The crisis continues. ACM SIGCUE Outlook, 1998, 26(2), 8–16 (1998)

Mifsud, L.: Alternative learning arenas - pedagogical challenges to mobile learning technology in education. In: Proceedings of the IEEE International Workshop on Wire-less and Mobile Technologies in Education (WMTE 2002), Växjö, Sweden, August 29-30, pp. 112–116 (2002)

MIT, MIT Project Oxygen: Pervasive Human-Centered Computing (2004), http://oxygen.lcs.mit.edu/

Mitchell, T.M.: Machine Learning. McGraw-Hill, Singapore (1997)

Moodle, Moodle (2009), http://moodle.org/about/

Moore, P., Hu, B.: A context framework with ontology for personalised and cooperative mobile learning. In: Shen, W., Luo, J., Lin, Z., Barthès, J.-P.A., Hao, Q. (eds.) CSCWD. LNCS, vol. 4402, pp. 727–738. Springer, Heidelberg (2007)

Moore, P., Hu, B., Wan, J.: Smart-Context: A Context Ontology for Pervasive Coopera-tive Applications. The Computer Journal, Advance Access (March 4, 2008), doi:10.1093/comjnl/bxm104

Moore, P.: The Complexity of Context in Mobile Information Systems. In: Proceedings of the 12th International Conference on Network-Based Information System (NiBS 2009), 1st International Workshop on Heterogeneous Environments and Technologies for Grid and P2P Systems (HETGP 2009), Indianapolis, USA, August 19-21, pp. 91–96 (2009)

Moore, P., Hu, B., Jackson, M., Wan, J.: 'Intelligent Context' for Personalised m-learning. In: Proc. of the International Conference on Complex, Intelligent and Software Intensive Systems (CISIS 2009), Fukuoka, Japan, March 16-19, pp. 247–254 (2009a)

Moore, P., Hu, B., Wan, J.: 'Intelligent Context' for Personalised Mobile Learning. In: Caballe, S., Xhafa, F., Daradoumis, T., Juan, A.A. (eds.) Architectures for Distributed and Complex M-Learning Systems: Applying Intelligent Technologies, Hershey, PA, USA. IGI Global (2009b)

Mungania, P.: The Seven E-learning Barriers Facing Employees (Tech. Rep.). University of Louisville, USA (2003)

Nardi, D., Brachman, R.J.: An introduction to description logics. Cambridge University Press, Cambridge (2002)

Papamarkos, G., Poulovassilis, A., Wood, P.Y.: Event-Condition-Action Rule Languages for the Semantic Web. In: Proc. of the Int. Workshop on Semantic Databases, pp. 309–327 (2003)

Pan, J., Paul, S., Jain, R.: MILSA: A Mobility and Multihoming Supporting Identifier Locator Split Architecture for Naming in the Next Generation Internet. In: Proc. of the IEEE Global Communications Conference (GLOBECOM 2008), New Orleans, LA, USA, November 30- December 4 (2008)

Paul, S., Jain, R., Pan, J., Bowman, M.: A Vision of the Next Generation Internet: A Policy Oriented Perspective. In: Proc of the British Computer Society (BCS) International Conferenceon Visions of Computer Science. Imperial College, London (2008)

Pop, D., Jichici, C., Negru, V.: A Combinative Method for Decision Tree Construction. In: Proc. of the Seventh International Symposium on Symbolic and Numerical Algorithms for Scientific Computing (SYNASC 2005), pp. 139–143 (2005)

Poslad, S., Laamanen, H., Malaka, R., Nick, A., Buckle, P., Zipf, A.: Crumpet: creation of user friendly mobile services personalised for tourism. 3G 2001. In: Proc. of the Second International Conference on 3G Mobile Communication Technologies, London (March 2001)

Rhodes, B.J.: The wearable remembrance agent: a system for augmented memory. In: Proc. of the 1st International Symposium on Wearable Computers, Cambridge, Massachusetts, USA, October 13-14, pp. 23–128 (1997)

Robson, C.: Real World Research, 2nd edn. Blackwell, Mass (2002)

Royer, E.M., Toh, C.-K.: A review of current routing protocols for ad-hoc mobile wire-less networks. IEEE Personal Communications 6(2), 46–55 (1999)

Rushby, N.: Where are the desks in the virtual class. ACM SIGCUE Outlook 26(2), 33–36 (1998)

Ryan, N., Pascoe, J., Morse, D.: Enhanced reality fieldwork: the context-aware archaeological assistant. In: Gaffney, Leusen, Exxon (eds.) Computer applications in archaeology. British Archaeological Reports, Oxford, UK (1997)

Salber, D., Dey, A.K., Abowd, G.D.: The Context Toolkit: aiding the development of context-enabled applications. In: Proc. of the ACM CHI 99 Human Factors in Computing Systems Conference (CHI 1999), Pittsburgh, Pennsylvania, May 15-20, pp. 434–441 (1999)

Scardamalia, M., Bereiter, C.: Knowledge building: theory, pedagogy, and technology. In: Sawyer, K. (ed.) Cambridge Handbook of the Learning Sciences, pp. 97–118. Cambridge University Press, New York (2006)

Senses, B.: Implementing Wireless Communication in Hospital environments with Bluetooth, 802.11b, and Other Technologies (Tech. Rep.). Medical Device & Diagnostic Industry (July 2003)

Shackle, G.L.S.: Decision, Order and Time in Human Affairs. Cambridge University Press, Cambridge (1961)

Singh, D., Abu Bakar, Z.: Wireless implementation of a mobile learning environment based on students' expectations. International Journal of Mobile Learning and Organisation 2007, I(2), 198–215 (2007)

W3C, W3C Semantic Web Activity (2009), http://www.w3.org/2001/sw/

Want, B.N., Schilit, N.L., Adams, R., Gold, K., Petersen, D., Goldberg, D., Ellis, J.R., Weiser, M.: An Overview of the PARCTAB ubiquitous Computing Environment. IEEE Personal Communications 2(6), 28–43 (1995)

Weiser, M.: The Computer for the 21st Century, September 1991, pp. 94–104. Scientific American (1991); reprinted in IEEE Pervasive Computing, pp. 19-25 (January-March 2002)

Weller, M.: Delivering learning on the net – the why, what and how of online education. Kogan Page, London (2002)

Wilkinson, K., Sayers, C., Kuno, H., Reynolds, D.: Efficient RDF Storage and Retrieval in Jena2, (Tech. Rep. HP Laboratories Technical Report HPL-2003-266) (2003)

Zadeh, L.A.: Fuzzy sets. Information and Control 8(3), 338–353 (1965)

Developing an Adaptive Learning Based Tourism Information System Using Ant Colony Metaphor

Soumya Banerjee[1], Monica Chis[2], and G.S. Dangayach[3]

[1] Department of Computer Science and Engineering, Birla Institute of Technology, International Center, Mauritius
[2] Siemens Program and System Engineering (Siemens PSE) Romania
[3] Department of Mechanical Engineering, Malviya National Institute of Technology, Jaipur, India

Abstract. Automation in learning process is one of the major technical breakthroughs in machine learning paradigm. A substantial boost in adaptive learning has been initiated by simple steps of bio-inspired algorithm to learn the collective pattern of tourist service environment. This chapter is devoted on a live project implementation and testing of a learning model prototype in tourist information system and service industry. The elaborated model is followed by result sessions, which demonstrate that artificial agents could mimic the collective service and product pattern effectively compared to other contemporary techniques. The cost optimization to address the service issues in tourism industry could also be achieved with the help of such prototype models.

1 Introduction

The development of searching technology has enabled rapid, nonlinear access to massive amounts of information on the Internet, thus helping to expand traditional learning in to a global context. At the present stage, searched contents are usually packaged and placed as learning objects in hypermedia educational tools for personalized enhancement and learning process. The information and learning is crucial for tourists as with remarkable web development in the recent years, a broad range of tourism information is distributed over millions of websites. This is a boon as well as a problem for the tourists. When information on a tour is needed, the user has to explore various web sites and collate discrete chunks of information [15]. Therefore, it leads to information overload and there is a necessity to develop a travel information and recommender system which could provide suitable recommendations to prospective travelers for tours and specifications (likely the cost, route, facility types, duration etc) they wish to undertake. When tourist want to get a touring recommendation, the adaptive learning based tourism information system looks for the particular tourist's group's selection patterns and prepare a suitable path for tourists in accordance with these patterns. The uses of groups' learning patterns are a kind of ant colony's intelligence, which can provide an impressive level of adaptability for other homogeneous tourists in this sort

F. Xhafa et al. (Eds.): Computational Intelligence for Tech. Enhanced Learning, SCI 273, pp. 59–77.
springerlink.com

of dynamic learning environment. Entire new group of tourists and their selection pattern could be explored as new search problem. The proposed chapter has emphasized the following crucial observations made on the developed prototype on the proposal:

- Adaptive learning has two features e.g. diversity and interactivity across the group members. The proposed model and prototype has revealed that intelligent learning technologies can make use of a group's behaviors to evaluate the current learning state and recommend the next best move or selection.
- A usability study was conducted using a prototype with eight users comprising beginner, intermediate as well as advanced users. All users are probable tourists.
- In developing the proposed adaptive learning based travel information system, learning style and repeated learning should be taken into account as the crisis points. This implies the different learning styles of individuals and group too. For an example, the learning could be conceptualized and memorized, depending on the level of learner. Considering this aspect the learning portal the proposed model could take the supporting contents those, who have the same learning style and then make it possible to tailor learning paths. Therefore, tour conductors will be able to rely on the resulting learning paths to further develop strategies and service policies for individualized tourists. Therefore, this model will be helpful for tourism informatics courseware development and training.
- The investigation in the model emphasizes the relationship of learning content to the learning style of each tourists or group of tourists in adaptive learning. An adaptive learning rule was developed in proposed *Travel Information System* to identify how learners of different learning styles may associate those contents which have the higher probability of being useful to form an optimal learning path.
- The model or similar tool is highly solicited for Mauritius and other neighboring countries to promote their tourism, which is one of the leading foreign income resources for them.

The complete chapter has been organized as follows:

Section 1 introduces the relevance and background of the problem following a brief motivational idea behind the project. Section 2 identifies the components of *Travel Information System* followed by its adaptability and intelligence. It also envisages similar motivation and other works in the context of tourism data mining. Section 3 discusses the core recent metaheuristic method (Ant Colony) for discovering group patterns that is designed to help tourists advance their on-line learning along an adaptive learning path. Section 4 discusses the implementation of the prototype followed by discussion and result in section 4.1 and 4.2 respectively. Section 5 describes the further extension of research on the developed tool.

1.1 Motivation of Using Bio-inspired Algorithm

The behavior of natural ants demonstrate the heuristics learning by following their shortest path while traveling for food from nest to food source. Their learning categorically distinguishes the optimal path using combinatorial algorithm and most importantly they accomplish their results with in a predominated time period. Mimicking the ants behavior inspires in the sense that the artificial ants can remember accepted path as well out of different alternatives. In tourism service, the tourists are having different combination of services available and based on certain constraints they accept or reject it. Hence, it is closer to certain combinatorial algorithm to decide instead of believing only on pure mathematical models. The flexibility of Ant Colony Optimization's parameters is another encouraging proposition to define and extend the model in the pretext of tourists learning pattern identification.

2 Components of Travel Information System

The *Travel Information System (TIS)* is one of the prime components for developing the *Travel Recommender System*. Like any other conventional information system, the development of efficient search functions is not an easy task in the field of information systems. As described in [1] this statement fits especially for the context of *TIS*, which differ from many other information systems in two special characteristics that are:

- TIS have to manage an extremely high amount of data, and in worst case extremely heterogeneous data.
- Users are not able to specify the search criteria exactly enough to receive from the system what they have in mind.

Therefore, searching becomes crucial to *Travel Information System* and subsequently queries and reply of tourists can be improved by augmenting the query relaxation techniques [2]. *Case Based Reasoning* methodologies have also added value in designing *TIS*, especially to solve a new problem by remembering a previous similar situation and by reusing its information and knowledge [3]. Steadily, the artifacts of *TIS* have been enhanced and the *TIS* are presented as a genuine advisor and recommendation system in the context of travel planning and management issues. Practically, *Travel Recommender Systems (TRS)* attempt to emulate offline travel agents by providing users with knowledgeable travel suggestions to facilitate their decision-making processes. Designing a recommender system application has some fundamental differences to software design for other applications.

The overall system architecture depends heavily on the choice of algorithms. The primary attributes of effective *Travel Recommendation System* are [4]:

- Database pre-selection - Quick reduction of large amounts of data.
- Content-based scoring - Allows to make use of typical tourist media information to add scores based on content and its semantics to the overall inference process.
- Incorporating implicit domain knowledge - Allows integrating handcrafted domain knowledge and expertise.

- Incorporating social aspects - Allows to integrate groups of users with a similar interest.
- User profiling - Representing users with respect the dynamic nature of users in tourism, as to combine all other approaches used in a system.

The attributes described above clearly demonstrate that, depending on the difference and diversification of tourist group the design of recommender system differs. The significant complexity has been observed to formulate and visualize the complete itinerary, of different tour containing information on destination, transportation, accommodation, entertainment, and attractions. Hence, emphasis has been given to build a tourism information system that overcomes this limitation and allows the tourist to enter the destination, choices of transport, and accommodation etc; and then view information about the entire tour as a continuous audio visual presentation [5] [6]. The advancement of web based presentation reinforce the trend of more realistic travel recommender system, but the learning phenomena of all those dynamic attributes of tourists in tourist information and recommender system could provide a different flair towards the development of more sophisticated and rapid decision support paradigm. The adaptive learning based tourism information system looks for the particular tourist's group's selection patterns and prepare a suitable path for tourists in accordance with these patterns.

The chapter is a suitable extension of tourists' group learning in the context of *Travel Information and Recommender System* to facilitate the learning process of diversified pattern in the selection of choice of services and interests for a homogeneous group of tourists. The adaptive and interactive behaviors of tourists of similar mode of interest can converge to fast recommendation process and design or development efforts both could be optimized. Even the risk and failure in tourism information system could be predetermined in different crisis points while learning the pattern. Therefore, learning and simulation based testing is extremely important for the prototype development of adaptive learning based *TIS*.

Agent and distributed object repository based models are the recent inclusions to achieve and intelligent *Visual Travel Recommender System* [7]. Recent research has revealed that intelligent learning technologies can make use of a group's behaviors to evaluate the current learning state and recommend the next best move. Several intelligent proposals have been developed: Dynamic learning recommendation [8], intelligent learning contents suggestion [9] and adaptive pedagogical path selection [10]. The use of a tourist groups' learning patterns are a kind of ant colony intelligence, which can provide an impressive level of adaptability for other homogeneous learners in this sort of dynamic learning environment [11]. From the viewpoint above, the idea of an adaptive learning path is to generate many useful patterns according to the shared learning contexts within which each learner learns and spend his/her time. Based on group and collective–learning patterns and employing an extended ant colony system approach, an *Adaptive Learning Based ant colony system (ALACS)* is proposed to construct a travel Information and Recommender model for finding suitable learning paths. This system uses ant colony's intelligence in a graph-based path structure that provides an effective method for solving most global optimal problems. Before preceding to next section that presents formal parameter and background of core Ant Colony(to tune up for adaptive leaning in tourism context) it will be worthy praise worthy to mention certain important breakthroughs on tourism data and information mining.

2.1 Tourism Data Mining: State of the Art

During the last decade, the quantity of potentially interesting products or information services available online has been growing rapidly and now exceeds human processing capabilities. Moreover, there are many information search situations where the users would like to choose among a set of alternative items or services, but do not have sufficient knowledge, capabilities or time to make such decisions. As such, there is a pressing need for intelligent systems that advise users while taking into account their personal needs and interests. There are substantial scopes associated with data mining approach in tourism sector. As mentioned in the previous section, the different wide ranges of categories could be derived by reciprocating with different tourism recommender system. One of the interesting results concerning the mining of tourism information system is the definition of *User Model (UM)* [16]. User modeling data integration can be achieved through a process that is referred to in this work as the *mediation* of *UM*s. Mediation of *UM*s is a process of importing and integrating the user modeling data collected by other recommender systems for the purposes of a specific recommendation task. Hence, the primary goal of the mediation is to instantiate *UM*s through inferring the required user modeling data from other data imported from other systems. The mediation enriches the existing *UM*s (or bootstraps empty *UM*s) in the target recommender systems and, as a result, facilitates provision of better recommendations.

This is the fundamental aspects of tourism recommender and information system to collect the relevant information and attribute from users and then gradually analyze it to infer certain decision. The data mining approach of tourism highlights the context sensitive of situation when tourist will ask for specific recommendation in a particular context [17]. The different important contexts are defined as follows:

- Environment context—captures the entities that surround the user
- Personal context—captures the state of the user and consists of two sub-components;
- the physiological context and the mental context;
- task context—captures what the persons (actors) are doing in this user context;
- social context—captures the social aspects of the current user context, such as friends, enemies, neighbors, co-workers and so on; and
- spatiotemporal context—captures aspects of the user
- Context relating to the time and spatial extent for the user context.

The relevance of context indicates the efficiency of different on line recommender system. Data mining studies identifies the reasons why established recommendation techniques like amazon.com cannot be directly applied to the tourism domain. Collaborative filtering techniques work best when there exists a broad user community and each user has already rated a significant number of items. As individual travel planning activities are typically much less frequent like, for example, book purchases, and in addition the items themselves may have a far more complex structure, it is hard to establish reasonable user profiles. Therefore, many approaches aim at eliciting the preferences and requirements in a conversational dialog (e.g., [18], [19]) using, for example, knowledge-based approaches for generating recommendations. Online users may be different with respect to their background knowledge, their mental models [19], or their capabilities of expressing their needs and requirements. Dialog design, usability aspects, and adaptivity are thus central in application and user interface design [20]. In [21] for instance, a critique-based dialog style is proposed which has already been successfully employed in other domains; [12] describes a tourism advisory application based on knowledge-based personalization and multi-step, adaptive dialogs. The problem of 'group recommendation' is another typical aspect in tourism-related recommender systems, i.e., the problem of generating proposals that 'maximize' the overall acceptance of members of a travel group that have different interests. Although this problem is not new in recommender systems (think of TV program or movie recommenders), there is only little research in the specific context of recommender systems in tourism [22].

Hence, the context, designing of user model is one of few emerging issues sin tourism data mining that should be addressed. The resent work integrates the tourist's pattern and learning of attribute in specific context.

3 Ant Colony Optimization: Group Pattern and Learning Components

Ant Colony Optimization (ACO), which is a specialized group of algorithms, is a recently developed population-based approach. It has been successfully applied to several NP-hard combinatorial optimization problems [11] [12]. As the name suggests, *ACO* was inspired by the observation of real ants' foraging behavior. Ants live in colonies. They use a cooperative method to search for food. While moving, ants initially explore the area surrounding their nest in a random manner. They initially leave a chemical pheromone trail on the ground. All ants can smell pheromone to choose their way. It implies that, they tend to choose the paths marked by strong pheromone concentrations. During the return trip, the quantity of pheromone that an ant leaves on the ground may depend on the quantity and quality of the food. At the same time, these pheromone trails progressively decrease by evaporation with time elapsing. This result in the amount of pheromone becomes larger on a shorter path. Then, the probability that an ant selects this shorter path is higher. One model of learning, that exhibits these features is the pheromone mechanism used by insects to guide their collective decision processes. Different variants of *ACO* have following basic parts: Solution construction, Heuristic information, Pheromone update and Local search. Out of these four components the relevant part is referred to the proposed model of algorithm, e.g. Pheromone update and local search.

Pheromone update: After an ant has completed its tour, the pheromone deposited on the arcs of that tour is used as the local updating rule. Firstly, the pheromone trail associated to each arc(r, s), is decreased by an evaporation factor as shown as the following:

$$\tau_{rs} \leftarrow (1-\rho)\tau_{rs} \tag{A}$$

Where, $\rho \in (0,1)$ is the evaporation rate. Secondly, each ant retraces the path it has followed and deposits an amount of pheromone $\Delta\tau_{rs}$ on each traversed connection; the local updating rule is defined as:

$$\tau_{rs} \leftarrow \tau_{rs} + \Delta\tau_{rs} \tag{B}$$

Local search: Let k be an ant located at node n, q is a random number (to be subjected to the combination of product and service pattern applicable for each probable tourist in the simulation experiment) uniformly distributed in interval [0, 1], and $q_0 \in [0, 1]$ is a parameter that determines the relative importance of exploitation versus exploration. The next node s is randomly chosen according to the following probability distribution. When $q \leq q0$, ant k at node n selects the next node s to move to if $q \leq q0$,

$$p_{ns}^{k} = \begin{cases} 1, & \text{if } s = \arg\max\limits_{u \in N_k(n)} \{\tau_{nu} \cdot \eta_{nu}^{\beta}\} \\ 0, & \text{otherwise} \end{cases}$$

else

$$p_{ns}^{k} = \begin{cases} \dfrac{|\tau_{ns}|^{\alpha}|\eta_{ns}|^{\beta}}{\sum\limits_{\mu \in N_n^k} |\tau_{n\mu}|^{\alpha}|\eta_{n\mu}|^{\beta}}, & \text{if } s \in N_k(n) \\ 0, & \text{otherwise} \end{cases} \tag{1}$$

where $N_k(n)$ denoted a set of nodes that remain to be traveled by ant k on node n. and regarding parameters α and β, their role is determined the level of consideration of the pheromone trails which is given a $\alpha \geq \beta$ or, with better heuristic preference, has a higher probability of being used in building a tour. Considering the core components of ACO, the mapping components for tourism are identified as:

- State Space for ant in each tour scenario

At each time step, each ant agent a is in some state s_a chosen from a known, finite set S of size K. We often think of K as being large, and thus want algorithms, whose running time scales polynomially in K and other parameters. We view s_a as the *action* taken by ant i in response to the recent population behavior. The collaborative action vector for tourist may be required iff the combination of different choices of services are made under same tour scenario.

- Initial State Distribution

In this model initial population state of ant agents $\vec{a}^0$ according to fixed but unknown P over S^a. During the learning the overall trajectory behavior of the tourist group is available to the particular instance.

- Tourist Policy Class

Each ant agent's strategy is drawn from a known class C of (typically probabilistic) mappings from the recent collective behavior into the ant's next state or action in S (s_a).

The above outline assists the model to envisage the learning model, objective function and the functional algorithm.

3.1 Selection of Learning Model

For learning *episodic* or *reset* model has been chosen. In this particular model, the learner has the luxury of repeatedly observing the population behavior from random initial conditions. This is mainly because the tourist choice or interest largely varies with respect to time and additionally it is most appropriate in (partially) controlled, experimental settings [13] where such "population resets" can be implemented or imposed.

The goal of the learner is to find a *generative model* that can efficiently produce trajectories from a distribution that is arbitrarily close to that generated by the true population. This generative model is able to tackle the multiple knowledge source of tour reacted information and also can access the contextual preferences of different tourists. Let L ($\vec{a}^0$, t) be a learning model, where $\vec{a}^0$ is the start initial state and G is the distribution generated by L over P. Similarly Q_f denotes the distribution of information in the tourist's learning G ($\vec{a}^0$, t) and thus the distance between (G, Q_f) should be minimum or small. The important parameter in this context might be the calculation of Euclidian Distance (ED) depending upon the combination of services and formation of similar group of tourists, who are soliciting those services.

$$\mathbf{ED}(\mathbf{T_1},\mathbf{T_2}) = {}^{\mathbf{n}}\sqrt{\sum_{\mathbf{i=1}} (S_1(T_1) - S_1(T_2))^2} \tag{2}$$

where $\mathbf{S_1}$ represents the value for a combination of different tourism related service options available for a tourist T_1(or T_2) and n represents the number of attributes. The evaluation of optimized service combination can be obtained (discussed in next section) and formation of the similar and homogeneous group of tourists is accomplished through ant colony based graph structure.

The representation form, we used is based on the idea of ordering tourists comparable to the classic traveling salesman problem. The first T_1 tourists belong to the first group who like some combination of services, the second T_2 tourists to the

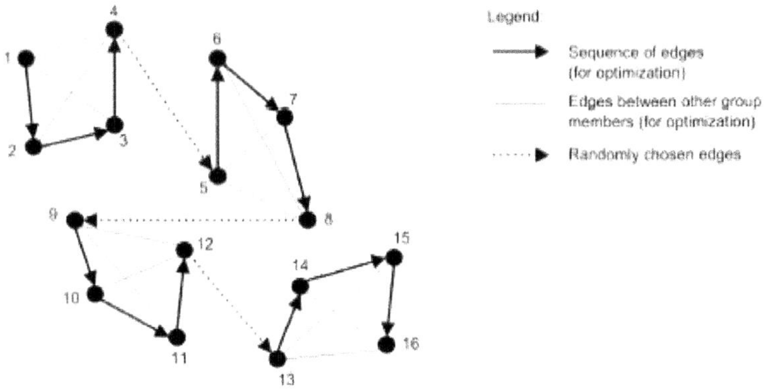

Fig. 1. Representation of the Grouping Problem as Graph (group size = 4)

second group and so on, whereby T is the maximum number of tourists per group. Figure 1 show this representation for a group size of four tourists, whereby the order is indicated by arrows. The principle is inculpated such that edges are used for pheromones and therefore indicate “goodness” of the edge within a group each newly assigned group member is linked not only to the last assigned group member but also to all other members of the group (indicated by solid lines in Fig.1). This is because the important information for optimization is not the order in which the tourists are assigned to a group but the fact that exactly these T students belong together. Therefore, also the decision to demark the particular tourist (based on service combination he liked) belong to a new group is performed randomly.

4 Prototype Implementation

Tourists queries are defined by a product and service pattern $q = q_1 . . . q_n$, where q_i is either 1 or 0, i = 1, . . . , n. If qi = 1, the user is interested in products that have the ith feature; if $q_i = 0$, the user has not (yet) declared any special interest in the ith feature. As a modeling standard Products are described by a fixed number of n Boolean features. For example, one standard hotel may have: a swimming pool, a gymnasium, multi- shopping facility, multi-cuisine, parking slot etc. Hence, each features could be treated as sub-product category and to be represented by a fixed-length string of bits, $b = b_1, . . . ,bn$, where bi = 1 means that the product has the ith feature and bi = 0 means that it does not have the feature. Tourists queries are defined by a product and service pattern $q = q_1 . . . q_n$, where q_i is either 1 or 0, i = 1, . . . , n. If qi = 1, the user is interested in products that have the ith feature; if $q_i = 0$, the user has not (yet) declared any special interest in the ith feature. Suppose, the core request made by the probable tourist learner is 1100(which invokes the combination service and features).As a modeling standard except the request 1100, all other possible combinations like 1110, 1101 and 0111 may also become possible. In order to find the optimal combination of product and service, where the similarity and distance of two or more distinct service plans P_1 and P_2, we incorporate the distance evaluation process (Refer Appendix B for details).

The implementation follows two different architectures; firstly an evaluation of combination of cluster of services for the tourists is applied for selecting the best next choice available and then algorithm initiates the learning session.

Products and services are described by a fixed number of n Boolean features. For example, one standard hotel may have: a swimming pool, a gymnasium, multi- shopping facility, multi-cuisine, parking slot etc. Hence, each features could be treated as sub-product category and to be represented by a fixed-length string of bits, $b = b_1, . . . ,b_n$, where $b_i = 1$ means that the product has the ith feature and bi = 0 means that it does not have the feature.

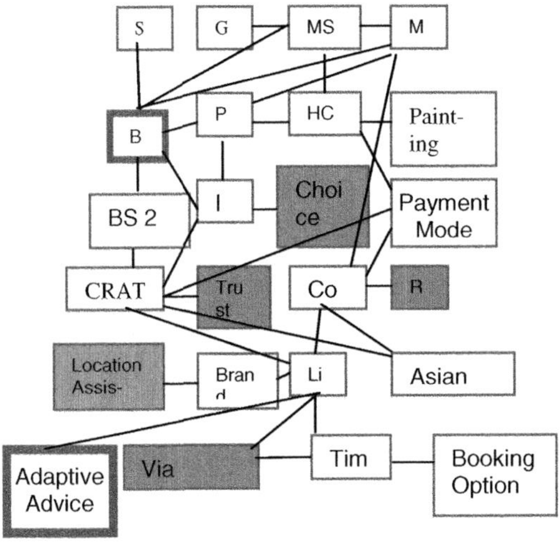

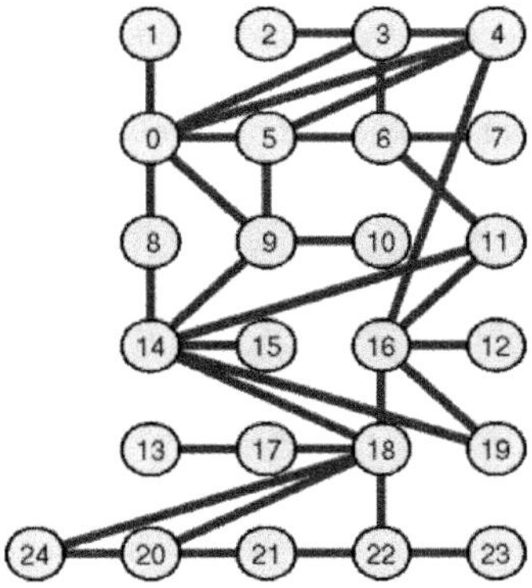

Fig. 2. Product and Service Combination for Tourist as Connected Graph

Table 1. Product and Service Combinational Data Vector (the presence and absence is denoted as 1 or 0 respectively)

Service/Product Option Vector	Semantics	Boolean Data Model	Weight as Feature
SP	Swimming Pool available	10101001011	0.6,0.2,0.5,0.4,0.3
GY	Gymnasium	01010001110	.
MS	Multi Shopping	01101101111	.
MC	Multi Cuisine	10110100011	.
RS	Room Service	01100101101	.
PS	Parking Slot	11010101101	.
IA	Interpreter Available	11010010001	.
Conti	Continental	01011100111	.
C.RATE	Concession Rate	10011001001	.
BS	Budget Shopping	01101101101	.
OF	Other Factor	11110011111	0.9.0.89, 0.9,1,0.98

4.1 High-Level Description of Prototype

The following is the list of initial parameters to implement the learning of tourist's service pattern for a particular instance, keeping the learning pre-defined:

T: Group of probable Tourist learners with different learning ability to choose the optimum service pattern.

Service feature-selection method to be evaluated and learned (under learning model and goal ($L(\vec{a}^0, t)$ and $G(\vec{a}^0, t)$ respectively Refer Section 3.1)

n_c: number of constraints in the initial query session after learning

$p = (p_1 \ldots p_n)$: user acceptance probabilities of service features

n_f : number of features suggested by service selection feature method at each learning sessions.

Recursive-iterate (R_i): Recursion and iteration id different processes, but after initial session if true then the system shows n_f additional features to learn when none of the previous n_f were accepted.

Based on the parameters the steps of implementation are as follows:

Step 1. Learning model $L(\vec{a}^0, t)$ is set to achieve $G(\vec{a}^0, t)$

Step 2. Choose n_c and n_f for combinations of service sequence with service pattern $Q \in \{0.........q-1)$ at random by Tourist T_1-T_8.

Step 3. If action of ant colony for next state S (s_a) and Let C be a class of probabilistic mappings from an input $\vec{\mathbf{x}} \in \mathbf{X}$ to an output $y \in Y$ where Y is a finite set for the colony distribution, we say that C is *polynomially learnable* if there exists an ant algorithm A such that for any $c \in C$ and any distribution D over X, if A is given access to a pattern of service producing pairs or services available.

Step 4. Calculate the Euclidian Distance (ED) according to eq. (2).

Step 5. Return Similar service pattern and optimum ED.

Step 6. Consider a new sequence of services $q = q_1, \ldots, q_n$ with $b = b_1, \ldots,$ bn, returned by initial query session. By the union bound, with probability $1 - \delta$, the probability that there exists any agent i and any $Q \in \{0.........q-1)$ such that the ED is optimum.

Step 7. If this is not the case return value of Step 6 for all i and $t \neq \in$ then the same sequence of states would have been reached if we had instead started at any other initial state S (s_a), where value of ED is tested.

Step 8. Set the pheromone value of ants interaction and epoch of learning; the pheromone evaporation factor has been modified as $1 - \delta$, and represented as a recursive-iterate (R_i) procedure (similar to [14]

Procedure daemon_actions

Let $k = current\_user\_style$ for $k = 1,\ldots,4$,

$R_{ij}^{k,style} = \{R^k\}$, and $R_{ij}^{k\ rank} = \{R\} - \{R^k\}$;

$\Delta\tau_{ij}^{k,style}(t) = \sum_{i=0}^{\upsilon} (\upsilon - n) \times Q \times R_{ij}^{k,style}(n)$,

where n is the index and υ is the total number of $R_{ij}^{k,style}(n)$ on x_{ij}

$\Delta\tau_{ij}^{k,rank}(t) = \sum_{n=1}^{\upsilon} Q \times R_{ij}^{k,rank}(n)$,

where n is the index, σ is the total number of $R_{ij}^{k,rank}(n)$ on x_{ij}, and $k = 1,\ldots,4$

$\tau_{ij}^{k}(t) = \rho\tau_{ij}(t-1) + \Delta\tau_{ij}^{k,style}(t) + \Delta\tau_{ij}^{k,rank}(t)$

end Procedure

Where,

R_{ij} The ratio of learning memory retention on arc (i, j)

$R_{ij}^{k,\ style}$ Learning ratio of a homogeneous ant which have the style k on arc (i, j)

$R_{ij}^{k,\ rank}$ Learning ratio of the ant with the style k on arc (i, j)

Step 9. C is still probable class of learners (tourists) derived from recursive-iterate process of different learning styles. Therefore, the difference in group of learner can be recorded.

Step 10. Stop the process and cluster the same group of tourist and made available that combination of product and services.

For detailed high level description refer **Appendix A**.

4.2 Result

The ratio of learning memory retention can be expressed as:

$$R_{ij} = e^{-(c\frac{t_{ij}}{s})^2} \tag{3}$$

Learning rate is obtained (R = 0.991752, changing the polarity from"-" to "+" signifies the positive expression on linear learning curve with predefined value of function f as shown in fig. 3) using Eq. (3), the results show that learning repetition is just 70% by the next day as shown in Fig.3. In fact, the forgetting curve of a learner falls 1-2% after 3–6 days. Therefore, a fixed learning constant is possibly unable to compensate for the actual learning situation. We may improve the learning constant C by regarding it as a variable value according to the accumulated time t so that, in the initial period, the fall in the learning constant c is smaller, and it increases along with time. This result in a more gradual and changing of the pheromone distribution over, polynomially learnable constant C, which could keep minimum value of Euclidean Distance(ED). It is demonstrated from simulation

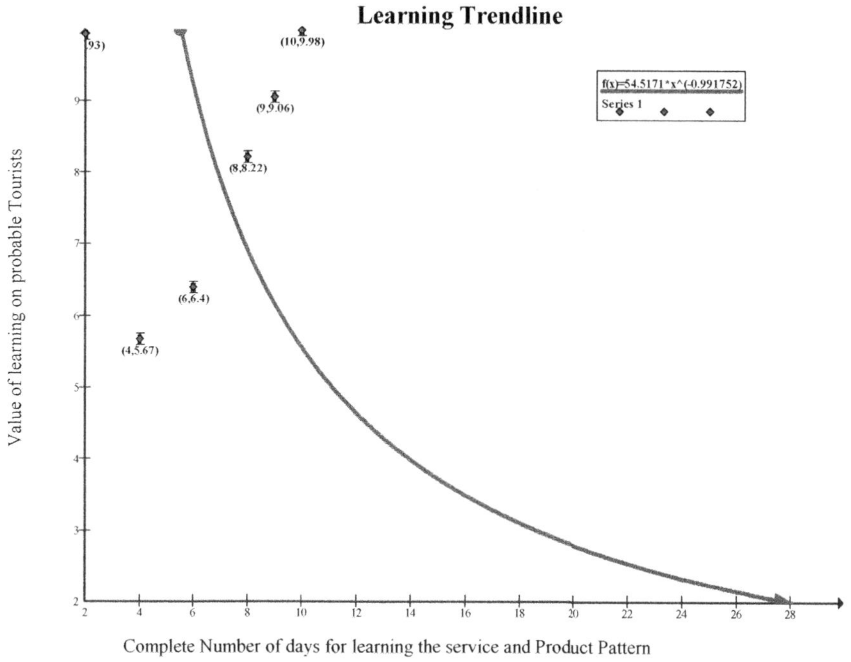

Fig. 3. Trend- line Analysis of Bio-inspired learning

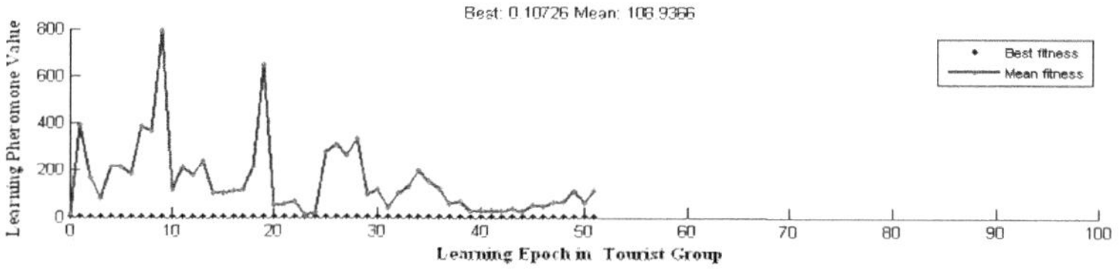

Fig. 4. Learning Pheromone versus Epoch in Tourist Group

optimum learning of combination of best available service pattern is converging approximately on 12-14 days (refer Fig. 3) of learning by the probable tourists.

Investigating the further learning trend and the relevance of pheromone learning, the simulation experienced (0-800) unit of pheromone deposition could yield peak earning in first 10-20 iterations or epoch (Refer Fig.4). Best learning value is found at 0.107026 unit of both axis overlap. This is significant in terms of the NP (Non Polynomial) hard nature of the problem as mentioned in table 1 in section 4. Therefore, the concept of local search for finding optimum learning pattern also becomes crucial, which has minimum value of Euclidian Distance and most suitable learning style adopted by the probable tourists. The learning style difference (due to beginner and advance tourist) of each one of the tourist learner produces different value path to find the optimum service pattern suitable for them and even for heir same choice of groups (shown in Fig. 5).

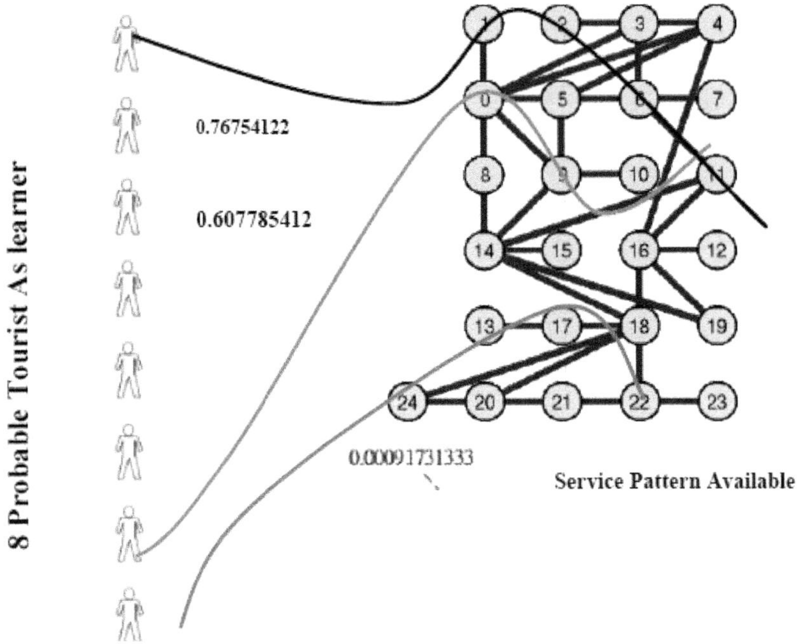

Fig. 5. Probable Tourists Learners interacts with Service pattern

The result of different data set according to the style and Euclidian Distance has been formulated with the help of prototype model and presented in the following table:

Table 2. Result and Comparison of 8 Probable Tourist learners

Learner	Learning Style Predefined Beginner = 1 Advanced = 4	From Service Pattern Node (0-24) Rank	Average ED	Standard Deviation Fitness of learning
1	3	2	361.434	0.031
2	1	3	1232.43	0.039
3	4	1	354.765	0.027
4	3	2	359.065	0.033
5	2	1	374.545	0.034
6	2	2	373.871	0.033
7	3	1	360.43	0.0271
8	4	1	354.713	0.028

Comparing the result of the experiment with 8 probable tourist learners with the result of the simple iterative algorithm in [13], aimed at finding learning heterogeneous groups according to the most plausible case of heterogeneity, it can be seen that the proposed algorithm delivers much better results. It has been observed that the similar learning style cluster is able to more nearer convergence like learner 1, 4 and 7 shares the equal capability of learning style of 3 categories(refer shaded portion of table 2). Regarding ED, the proposed algorithm includes the ED values of all combinations of group members. Nevertheless, the average ED values of the proposed algorithm are slightly higher, which indicated a much better heterogeneity of the style and learning pattern. The available combination of pattern is complex and recursive-iterate in nature (for sample question refer Appendix B).

5 Conclusion

In this chapter an adaptive learning based bio-inspired algorithm is presented in the context of Travel Information System. Adaptive learning has two features e.g. diversity and interactivity across the group members. The proposed model and prototype has revealed that intelligent learning technologies (of ant agents) can make use of a group's behaviors to evaluate the current learning state and recommend the next best move or selection. The subject of learning is to optimize the different available service and product pattern across the heterogeneous group of tourists simulating the pheromone deposition and update mechanism of ants. The learning ability of the tourists is evaluated depending on their familiarity of contents and competency to select best service pattern and thus cluster of tourists pertaining to same pattern can be identified. This phenomenon facilitates better service prospect and less searching cost on travel portal as well.

Acknowledgments

The present case study and data has been supported by *Department of Tourism, Republic of Mauritius.* The work is partially supported by *Department of Computer Science, Birla Institute of Technology, International Center, Mauritius, and Mauritius Research Council, Govt. of Mauritius.*

References

[1] Palkoska, J., Pühretmair, F., Tjoa, A.M., Wagner, R., Wöß, W.: Advanced Query Mechanisms in Tourism Information Systems. In: Proceedings of the International Conference on Information and Communication Technologies in Tourism (ENTER 2002), pp. 438–447. Springer, Innsbruck (2002)

[2] Bridge, D., Göker, M., McGinty, L., Smyth, B.: Case-based recommender systems. Knowledge Engineering Review 20(3), 315–320 (2006)

[3] Adomavicius, G., Tuzhilin, A.: Toward the Next Generation of Recommender Systems: A Survey of the State-of-the-Art and Possible Extensions. IEEE Trans. Knowledge and Data Eng. 17(6), 734–749 (2005)

[4] Felfernig, A., Friedrich, G., Jannach, D., Zanker, M.: An Environment for the Development of Knowledge-based Recommender Applications. International Journal of Electronic Commerce (2007)

[5] Ponnada, M., Jakkilinki, R., Sharda, N.: Tourism recommender systems: Current technology and future directions. In: Pease, W., Rowe, M., Cooper, M. (eds.) Information and Communication Technologies in support of the tourism industry. Idea Group Inc., Hershey (2006)

[6] Ponnada, M., Sharda, N.: A High level model for developing Intelligent Visual Travel Recommender Systems. In: Sigala, M., Mich, L., Murphy, J. (eds.) ENTER 2007: 14th annual conference of IFITT, the International Federation for IT & Travel and Tourism, Ljubljana, Slovenia, January 24-26. Springer, Vienna (2007)

[7] Ponnada, M., Sharda, N.: A High level model for developing Intelligent Visual Travel Recommender Systems. In: Sigala, M., Mich, L., Murphy, J. (eds.) ENTER 2007: 14th annual conference of IFITT, the International Federation for IT & Travel and Tourism, Ljubljana, Slovenia, January 24-26. Springer, Vienna (2007)

[8] Chen, J.N., Huang, Y.M., Chu, W.C.: Applying Dynamic Fuzzy Petri Net to Web Learning System. Interactive Learning Environments 13(3), 159–178 (2005)

[9] Huang, Y.M., Chen, J.N., Kuo, Y., Jeng, Y.L.: An intelligent human-expert forum System based on Fuzzy Information Retrieval Technique. Expert Systems with Applications 34(2) (2007)

[10] Semet, Y., Lutton, E., Collet, P.: Ant colony optimization for e-learning: Observing the emergence of pedagogic suggestions. In: IEEE Swarm Intelligence Symposium, pp. 46–52 (2003)

[11] Dorigo, M., Birattari, M., Stiitzle, T.: Ant colony optimization: Artificial Ants as a Computational Intelligence Technique. IEEE Computational Intelligence Magazine 1(4) (2006)

[12] Dorigo, M., Gambardella, L.M., Birattari, M., Martinoli, A., Poli, R., Stützle, T. (eds.): ANTS 2006. LNCS, vol. 4150. Springer, Heidelberg (2006)

[13] Bekele, R.: Computer Assisted Learner Group Formation Based on Personality Traits. Ph.D Dissertation, University of Hamburg, Hamburg, Germany (2005), `http://www.sub.unihamburg.de/opus/volltexte/2006/2759` (Retrieved February 10, 2009)

[14] Doerr, B., Neumann, F., Sudholt, D., Witt, C.: On the runtime analysis of the 1-ANT ACO algorithm. In: GECCO 2007: Proceedings of the 9th annual conference on Genetic and evolutionary computation, pp. 33–40. ACM, New York (2007)

[15] Venkataiah, S., Sharda, N., Ponnada, M.: A Comparative Study of Continuous and Discrete Visualization of Tourism Information. In: Proceedings of the International Conference on Information and Communication Technologies in Tourism, ENTER 2008, Innsbruck, Austria, January 23-25 (2008)

[16] Adomavicius, G., Sankaranarayanan, R., Sen, S., Tuzhilin, A.: Incorporating contextual information in recommender systems using a multidimensional approach. ACM Trans. Inform. Syst. 23(1), 103–145 (2005)

[17] Goldberg, K., Roeder, T., Gupta, D., Perkins, C.: Eigentaste: a constant time collaborative filtering algorithm. Inform. Retr. 4(2), 133–151 (2001)

[18] Ricci, F., Del Missier, F.: Supporting Travel Decision Making through Personalized Recommendation. In: Karat, C.M., Blom, J., Karat, J. (eds.) Designing Personalized User Experiences for E-Commerce, pp. 221–251. Kluwer Academic Publisher, Dordrecht (2004)

[19] Xiang, Z., Fesenmaier, D.: An analysis of two search engine interface metaphors for trip planning. Information Technology & Tourism 7(2), 103–117 (2005)
[20] Ricci, F.: Travel recommender Systems. IEEE Intelligent Systems, 55–57 (November/December 2002)
[21] Ricci, F., Nguyen, Q.N.: Critique-Based Mobile Recommender Systems. OEGAI Journal 24(4) (2005)
[22] Ardissono, L., Goy, A., Petrone, G., Segnan, M., Torasso, P.: INTRIGUE: personalized recommendation of tourist attractions for desktop and handset devices. Applied AI, Special Issue on Artificial Intelligence for Cultural Heritage and Digital Libraries 17(8-9), 687–714 (2003)

Appendix -A

High Level Description of the proposed Learning Algorithm for Tourist Service pattern (referred in Section 4.1)

LearningTouristServcePattern(S, n_c, p, L ($\vec{a}^0$, t), G ($\vec{a}^0$, t), n_f, Recursive-iterate -R_i)

```
for each s ∈ S do
    q ← InitialQueryStringforlearning(x, n_c) % build the initial query
    loop ← true
    interaction ← 0
    features ← true
    while count (q) > 50 AND Service features do
  sf ← L ( a⃗^0, t) [q] % sort features not yet constrained in q with method of
L ( a⃗^0, t),
        j ← 1
      while loop AND j ≤ sf.size do
            xi ← sfj % the j–th feature in s_f
            if accept (x_i, p_i, s) then
                q ← q ∪ {xi = si} // Apply ED formula eq. (2)
                Pheromone  interaction ← interaction + 1
                loop ← false % a new constraint has been found then exits while
            else
                if (j modulo n_f = 0) then
                    Pheromone_interaction ← interaction + 1
                        if NOT Recursive- iterate (R_i) then
```

```
                        loop ← false % not consider other n_f features
                        features ← false % exit also from main while
                    end if
                      end if
                          j ← j + 1
                    end if
                end while
                   if j = sf.size + 1 then % q has not been updated
                        features ← false % exit from outer while
                end if
                      end while
                end for
```

Each experiment consists of 20 runs. The parameter for ACS are assumed according to literature [6] or based on experiments. We assumed that, β =1, ρ =0.1, q_0=0.9, and the number of ants=10. Simulation is done through Borland C++ 5.0 language using 2.53 GHz Pentium IV processor with 1 GB of RAM running the standard Windows XP Professional operating system.

Appendix –B

Sample experiment contexts for choosing optimum service pattern for 8 probable tourists used in simulation (referred in Section 4.2).

Table B1. Alteration and impact Feature Value for different tourism services

Sample Snap Rule	**Immediate Previous Product – Service Session (IPS)**	**Probable Combination Set 1**	**Probable Combination Set 2**
Service Plan Details P_1 and P_2	All Facilities, except the price range	All Facilities, except there is no Concession range	Smarter Service, more Costly Service plan
Price Range	=40,000 MUR[1]	> 60,000 MUR	More than 11,00,000 MUR

[1] MUR: Mauritian Rupees.

Table B1. (*continued*)

Brand	Not Certain	Moderate	True International Brand
Payment Mode	Flexible & Current Pay	Flexible & Current Pay	Advance/Post Payment
Polled/Chosen	Partial	Not Popular	Mostly chosen plan

The flowing sample context question have been sorted to collect feedback experience of 8 probable tourist learners while using the proposed plug-in tool in **Travel Information System (TIS)** and **Travel Recommender System (TRS)**.

Question: 1. How the combination of your liked service has been formed?

Question: 2. How fast and effective was it to retrieve information from TIS?

Question: 3. Whether the presentation on browser reduced your time to chose the best for your choice?

Question: 4. Would you like to find out your closer group of interest in terms of service selection?

Intelligent and Interactive Web-Based Tutoring System in Engineering Education: Reviews, Perspectives and Development

Arun S. Patil[1] and Ajith Abraham[2]

[1] CQUniversity Mackay Campus, Australia
a.patil@cqu.edu.au
[2] Machine Intelligence Research Labs (MIR Labs)
Scientific Network for Innovation and Research Excellence
P.O. Box 2259
Auburn, Washington 98071-2259, USA
ajith.abraham@ieee.org
http://www.mirlabs.org
http://www.softcomputing.net

Abstract. Due to the rapid growth of the use of computers and increasing use of the Internet in education a large number of Web-based educational applications have been developed and implemented. However, very few of them are pedagogically intelligent and interactive for learning purposes. The Web-based intelligent learning has become more effective in the past decade due to increasing use of the Internet in education. A literature search indicates that there is a lack of relevant comprehensive research concerning the efficiency of computer-assisted instructions used in engineering education. The main focus of the research described in this chapter is on the comprehensive review of design and development of the Web-based authoring tool for an Intelligent Tutoring System in engineering education. The chapter outlines and discusses important issues of the development of Intelligent Tutoring System (ITS) in engineering education with an example of the development of a Web-Based Computer-Assisted Tutorials and Laboratory Procedures (WCALP).

1 Computers in Education

The computer-assisted instruction (CAI) in education effectively began in the late 1950s with the use of time-sharing computers, where communication was mainly through the mainframe computers. Hativa (1989, p. 81) suggested that ... *the computer-assisted instruction (CAI) systems designed for the same educational objectives may be based on entirely different concepts and methods of operations and may produce different students attitudes and achievements*. According to Atkinson and Wilson (1969), the three important factors that contribute heavily to the growth of computer-assisted instruction are:

- Development of programmed instruction;
- Mushrooming of electronic data processing;
- Increasing aid to education sectors.

F. Xhafa et al. (Eds.): Computational Intelligence for Tech. Enhanced Learning, SCI 273, pp. 79–97.
springerlink.com

Kulik's research in the 1980s was focused on the use of computer-mediated instruction and found positive research outcomes. The meta-analysis, conducted on 254 controlled evaluation studies showed positive learning effect of computer-based instruction (Kulik & Kulik, 1991). Kulik's meta-analysis studies were carried out when software tutorial programs had limited options. Nevertheless, Kulik's work has influenced educationalists and developers of computer-based instruction authoring systems for the last two decades. In an interesting study into assessing the effects of simulations and higher order thinking technologies with over 13,000 fourth and eighth grade students, Wenglinsky found gains in students' mathematics scores (Schacter, 2003). Also, the higher order use of computers was affected positively on most students' academic achievements. According to Pudlowski (1995), the use of computers in the teaching process should be considered as a supplement to available methods of instruction, rather than the universal remedy for all of the deficiencies that occur in the computer-based education.

The three most common modes of CAI used in learning are:

- Tutorials;
- Drills and practices;
- Interactive simulations.

1.1 Advantages of CAI

Computer-assisted instruction plays a very important role in the modern education process and extensive research shows increasing evidence of the use of computers in the teaching-learning process. There is a body of evidence that computer-assisted instruction, which focuses on higher-order learning in technical education, has been more effective than traditional instruction (Yaakub, 2001). It has been shown that computer-assisted instruction has many advantages in teaching-learning processes, including the following important elements:

- Learners can progress at their own rate and pace.
- Learners can have more individual considerations.
- Immediate feedback is available.
- Learner's response can be recorded.
- Visualisation is properly increased.
- Remedial teaching is eliminated and the problem of discontinuance of learning is avoided.
- A deeper analysis of transfer phenomena is permitted and facilitated.
- An opportunity for the introduction of more effective methods for testing in the teaching/learning process is provided.
- Qualitative and quantitative analysis of student achievements is facilitated.
- Access to supplementary material through links is provided.
- Several complex process and interactions can be explained in simple format.

However, comprehensive research is required to determine the best methodology to be applied to the design and development of computer-assisted instruction, as well as the efficiency of the teaching/learning processes based on this particular method of instruction (Patil, 2004).

2 Computer Assisted Authoring in Engineering Education

There has been significant growth in the use of computers in engineering education; this varies from classroom learning to remote wireless access distance learning. The most important and core part of computer-assisted instruction is to devise and design teaching algorithms. In order to develop any computer-assisted authoring program, two important components to be considered by the developers are:

1. High-performance software;
2. Accessibility.

Advancements in sophisticated computer software tools and hardware technology have resulted in significant advances in courseware authoring tools. The use of modelling and simulation techniques in engineering education is becoming increasingly common over recent years. Early CAI programs were very simple with minimal interaction required. However the use of spreadsheets expanded considerably in the late 1980s and, since then, they are commonly used in almost all engineering disciplines. Advantages of using spreadsheet include the fact that they are easy to learn, easy to debug, and have appropriate graphic and scientific functions. Their excellent problem-solving speed made it especially popular in engineering education (Wankat & Oreovicz, 1993). The development of suitable software was somewhat complex due to the high degree of iteration, as well as problems with debugging. As a result, the availability of authoring tools was limited due to constraints with hardware platforms (Toogod & Wong, 1993).

Since engineering education involves calculations and problem solving, symbolic algebra programs were designed and developed, such as *Mathematica*, *Derive*, *Mapple*, etc. However, in the last decade, parallel development had been evidenced with regard to hardware and software. This has occurred at an incredible rate and resulted in the development of commercial software applications like *SPICE*, *pSPICE*, *ASPEN*, *LabVIEW*, *Labtech*, *WebCT*, *ToolBook* and various CAD programs. Considering the common use of computer operating systems, such as Microsoft Windows, dedicated CAI authoring packages like *Authorware Professional for Windows* have been developed as new generation software packages (Toogood & Wong, 1993). However, the various easily available commercial packages available in the marketplace can be very complex to use, expensive and lack those functions that are considered important for certain learning tasks (Ertugrul, 2000).

3 Intelligent Tutoring Systems (ITS)

An Intelligent Tutoring Systems (ITS) has its roots in the generative Computer-assisted instruction (CAI) and is often known as Intelligent Computer-Assisted Instruction (ICAI) (Sleeman & Brown, 1982). Computer-aided instruction were the forbears to ITS, which started in the late 1950s. The more sophisticated tutoring systems were developed due to the common use of artificial intelligence, which basically started in the 1970s. The intelligent CAI system has the potential to provide the necessary interaction to suit various learners, and differentiates clearly between Computer-Assisted Instruction (CAI) and Intelligent Computer-assisted instruction (ICAI) (Boulay, 1998). Table 1 briefly lists the main differences between CAI and Intelligent CAI.

Table 1. Guide to differentiating intelligent CAI systems from CAI systems

Computer-assisted instructional systems (CAI)	Intelligent computer-assisted instructional systems (ICAI)
Knowledgeable	Highly knowledgeable
Can be intelligent	Maximum intelligence
Slow learning	Fast learning
Tutor controlled	Student controlled
Active feedback	Interactive feedback

The Intelligent Tutoring System is one of the fast developing and most popular areas of Computer-Assisted Learning (CAL). According to Ong and Ramachandran (2000), the beauty of ITS technology is that it provides learners with highly interactive learning environments that enable students to practice their skills by carrying out learning tasks. The use of intelligent machines for teaching in education has been around since 1926, when Sidney L. Pressey built a machine with multiple choice questions and answers, which also provided immediate feedback to the user (Thomas, 2003). The three essential components or models of ITS are as follows:

- Knowledge of the domain or expert model;
- Knowledge of the learner or student model;
- Knowledge of teacher strategies or instructor model (Thomas, 2003, Ong & Ramachandran, 2000).

3.1 Artificial Intelligence, Expert Systems and Simulations

The principle of AI made computers more useful, as well as intelligent, in order to utilise them in all the fields of human life. The application of AI principles is the next advanced step to a Web-based ITS, which began in the 1970s and 1980s. Since then, the influence of AI on software technology has considerably increased. As a result, the use of AI techniques in teaching/learning, such as expert systems, simulations and robotics, etc, has become a major factor in the development of Web-based intelligent authoring systems. AI is an advanced scientific technology that is used for efficient computer-based problem-solving techniques in various disciplines. The literature survey shows that the advancement in hardware and software systems has resulted in the development of numerous AI-based authoring tools with the help of sophisticated programming languages, such as LISP or Prolog. There are numerous interactive simulation tools developed in basic engineering disciplines, for example: a prototype of an intelligent case-based process planning system (Yang & Lu, 1993) and three-dimensional mechanical assemblies (Fang & Liou, 1993). However, most of these simulation systems are standalone in nature.

The important contribution of AI in computer-based education is to provide knowledge-based access to resources. Wilson and Welsh (1991) divided AI into three broad areas where knowledge-based systems or expert systems can have important implications for education and training. The history of computerised educational measurement system shows that each generation of educational measurement has shown an increased use of AI and expert systems approaches in order to improve educational measurement activities (Olsen, 1991). The four important generations highlighted by Olsen (1991) are:

- Computerised testing;
- Computerised adaptive testing;
- Continuous measurement;
- Intelligent measurement.

In engineering, science and technology subjects, most of the concepts comprise of mathematical calculations, complex phenomena and graphical representations. Students need to practise in order to familiarise and understand the concepts. Simulations and computer modelling are the most suitable techniques, which provide the proper learning environment. Wedekind (1988) outlined the common structure of all computer simulation programs used in computer-assisted learning and according to him computer simulation programs include input and output routines, plus numerical and graphical routines. ITSs also have the advantages of incorporating simulations, which help in exercising and enhancing learners' expertise in specific areas. SOPHIE, QUEST and STEAMER projects are a few examples of simulation-based tutoring systems. However, the special purposes simulation-based tutoring projects were developed in the late 1970s and 1980s. The sophisticated authoring systems such as; DIAG, RIDES, VIVIDS, etc, developed in the 1990s, provide essential graphical tools, so that students can interact with the software and use a variety of simulation techniques. These simulation techniques are easy to modify and maintain. An authoring system such as RIDES can also automatically generate instruction, which may include feedback and evaluation (Murray, 1997).

3.2 Working of an ITS

The three major components of any ITS coordinate and work together in order to produce better learning outcomes. These components are listed as below:

- Teacher strategies;
- Learner knowledge;
- Domain knowledge.

Although each component has its specific identity and functioning ability, they always interact and coordinate with each other that help in guiding learners and interfacing them to expert in the subject. In a typical formal teaching situation, guidance is provided based on the learner's performance observed and gained during the instruction. However, in ITS-based learning, highly interactive techniques, such as simulations, AI and expert systems are incorporated with the learning in order to perform the action. The information received by the learner is again used as a feedback to provide further instructions and the cycle is repeated. Beck et al (1996) categorised ITSs on two dimensions, namely: abstraction of the learning environment and the knowledge type of the instruction.

3.3 ITS and World Wide Web

Most of the traditionally developed ITSs use static media, such as CD-ROMs. However, it is very difficult to deploy the advancement and changes in the courseware in such systems; hence it is generally not suitable for basic learning strategies. The way to minimise the disadvantages inherent in traditional Intelligent Tutoring Systems is to utilise World Wide Web (WWW). Since its inception in 1990, the Internet (WWW or Web) has quickly emerged as a powerful new tool for connecting people and information on a global scale (Reed & Afjeh, 1998). The Web technology can also be utilised for various learning programs based on multimedia/hypermedia techniques. ITS developers mostly concentrate on designing a program that can provide effective instructions for the desired learning task. The developer's task has become more advanced and useful but challenging due to the common use of Internet technology in educational instructional systems. Recently, several advanced ITSs have been used for open and distance learning in many universities that provide learners with adequate sources of learning material for unlimited, unconditional use via the Internet.

4 ITS Developments

In the initial phase of ITS developments, several efforts have been undertaken to develop tutoring systems, mostly for teaching computer programming to students at the university level (Larkin and Chabay, 1992). However, recent research on ITS developments indicates that there are few dedicated ITSs developed in the engineering domain. Dear (1991) identified the trend in developing computer-assisted instruction regarding the relationship of the content and the code. According to him, in traditional authoring systems developed with general purpose programming languages (eg BASIC, COBOL, PASCAL, etc), the code and the contents were one and the same, yet this trend is changing with regard to AI-based authoring systems in which content is being separated from the code.

Curilem et al (2006) propose a mathematical model of Intelligent Tutoring Systems (ITS), based on observations of the behaviour of these systems. One of the most important problems of pedagogical software is to establish a common language between the knowledge areas involved in their development, basically pedagogical, computing and domain areas. Authors presented an example demonstrating how the formalization was used to design the adaptive mechanism of and to adapt its Interface Module to some student characteristics.

Crowley and Medvedeva (2006) described the development of a general intelligent tutoring system for teaching visual classification problem solving. The architecture incorporates aspects of cognitive tutoring system and knowledge-based system design within the framework of the unified problem-solving method description language component model. Based on the domain ontology, domain task ontology and case data, the abstract problem-solving methods of the expert model create a dynamic solution graph. Student interaction with the solution graph is filtered through an instructional layer, which is created by a second set of abstract problem-solving methods and pedagogic ontologies, in response to the current state of the student model.

Huang et al. (2006) proposed three kinds of learning parameter improvement mechanisms to establish effective parameters that are frequently used in the learning platforms. The proposed learning parameter improvement mechanisms can calculate the students' effective online learning time, extract the portion of a message in discussion section which is strongly related to the learning topics, and detect plagiarism in students' homework, respectively. The derived numeric parameters are then fed into a Support Vector Machine (SVM) classifier to predict each learner's performance in order to verify whether they mirror the student's studying behaviors. The experimental results show that the prediction rate for the SVM classifier can be increased up to 35.7% in average after the inputs to the classifier are "purified" by the learning parameter improvement mechanisms.

Woo et al. (2006) main research objective was to build an intelligent tutoring system capable of carrying on a natural language dialogue with a student who is solving a problem in physiology. Analysis of a corpus of 75 hour-long tutoring sessions carried on in keyboard-to-keyboard style by two professors of physiology tutoring first-year medical students provided the rules used in tutoring strategies and tactics, parsing, and text generation. The system presents the student with a perturbation to the blood pressure, asks for qualitative predictions of the changes produced in seven important cardiovascular variables, and then launches a dialogue to correct any errors and to probe for possible misconceptions. The natural language understanding component uses a cascade of finite-state machines.Results of experiments with pretests and posttests have shown that using the system for an hour produces significant learning gains and also that even this brief use improves the student's ability to solve problems more then reading textual material on the topic. Han et al. (2005) described an intelligent tutoring in a collaborative medical tutor for Problem Based Learning (PBL). Authors used Bayesian networks to model individual student clinical reasoning, as well as that of the group. The prototype system incorporates substantial domain knowledge in the areas of head injury, stroke and heart attack. Tutoring in PBL is particularly challenging since the tutor should provide as little guidance as possible while at the same time not allowing the students to get lost. From studies of PBL sessions at a local medical school, the authors identified and implemented eight commonly used hinting strategies.

Stathacopoulou et al. (2005) proposed a neural network implementation for a fuzzy logic-based model of the diagnostic process is proposed as a means to achieve accurate student diagnosis and updates of the student model in Intelligent Learning Environments. The neuro-fuzzy synergy allows the diagnostic model to some extent "imitate" teachers in diagnosing students' characteristics, and equips the intelligent learning environment with reasoning capabilities that can be further used to drive pedagogical decisions depending on the student learning style. The neuro-fuzzy implementation helps to encode both structured and non-structured teachers' knowledge: when teachers' reasoning is available and well defined, it can be encoded in the form of fuzzy rules; when teachers' reasoning is not well defined but is available through practical examples illustrating their experience, then the networks can be trained to represent this experience. Jaques and Vicari (2006) described the use of mental states approach, more specifically the belief-desire-intention (BDI) model, to implement the process of affective diagnosis in an educational environment. Authors used the psychological OCC model, which is based on the cognitive theory of emotions and is possible to be implemented computationally, in order to infer the learner's emotions from his actions in the system interface. The proposed architecture profit from the reasoning capacity of the BDI model in order to infer the student's appraisal (a cognitive evaluation of a person that elicits an emotion) which allows to deduce student's emotions. The system reasons about an emotion-generating situation and tries to infer the user's emotion by using the OCC model. Besides, the BDI model is very adequate to infer and also model student's affective states since the emotions have a dynamic nature.

Aberek and Popov (2004) presented an intelligent tutoring system STATFAG for education on design, optimisation and the manufacturing of gears and gearing. STATFAG serves several purposes. It helps lecturers to plan and execute optimal lessons for various subjects; it makes self-studies of students of higher level of education at the Faculty of Mechanical Engineering much more efficient; and also assists less experienced graduates who must independently design gear assemblies in companies. Hwang (2003) proposed a conceptual map model, which provides learning suggestions by analyzing the subject materials and test results. A testing and diagnostic system is also implemented on computer networks based on the novel approach. Experimental results have demonstrated that the novel approach benefits students and deserves further investigation.

5 A Case Study of WCALP

Woolf (1992) claims that several iterative cycles are required to develop an intelligent tutor or any AI system. It is very hard to design and develop an ideal ITS that may cater all pedagogical needs and also will suit with available software technologies. However, while developing any ITS developers must consider two important levels of integration: one from the developer's point of view and the other from the learner's point of view. Beck et al (1996) suggested a strategy for the design of effective ITSs with five essential components, namely: the student model, the pedagogical model, the domain knowledge, the communications model and the expert model. Their research on the applications of AI in education added the *expert model* as the fifth component to the four major components already identified by Woolf. Furthermore, they pointed out the importance of a large research potential in multiple teaching strategies in the pedagogical model. Table 2 gives a list of essential points to be considered to formulate the strategy in developing the Web-based educational system or a Web-based ITS.

Table 2. Essential points to be considered in designing an ITS strategy (Patil, 2004)

Developer's View	Learner's (End User's) View
• Course material • Systems available • Software design • Accessibility • Learning pedagogy • Learning outcomes • Cost-benefits	• Learning approach • Change of behaviour • Metacognition • Time and usability • Cost-benefits • Assessments and evaluations

5.1 Methodology

The development of Web-based ITS have two important levels of integration: one from the developer's point of view and the other from the learner's point of view. To develop any Web-based ITS is a tedious task. Moreover, the developers must keep in mind that the Web-based educational systems should have the following basic characteristics:

1. Dynamic;
2. Interactive;
3. Intelligent;
4. Adaptive;
5. Flexible;
6. Learner centred (Patil, 2004).

It is important to have proper organisation of knowledge or curriculum to be taught in designing the Web-based ITS. Albacete and VanLehn (2000) identify the manner in which the desired knowledge is presented is very important factor in human tutoring techniques. Nevertheless, CAI, or Web-based learning, is an evolutionary process and has its own unique characteristics regarding the choice of appropriate pedagogy. In designing the project pedagogy, the following factors will be taken into consideration:

- Content of the course
- Learner's activities
- Mode and level of interaction
- Impact of new approach
- Performance indicator
- Learning outcomes.

5.2 System Architecture

Any computer-assisted learning or ITS must have essential courseware design that can easily fulfil the important pedagogical requirements in the learning strategy. It has been proved with much research in the area of online learning that an integral approach to online learning can be used to promote students' critical use, understanding and application of materials. The proposed system architecture of a Web-based pedagogical ITS module can be explained with a block diagram shown in Figure 1. The most important part of such an architecture is a domain layer with learners' interaction within the pedagogical control unit and their responses. It is also possible to enhance the pedagogical aspects in the teaching-learning environment by incorporating pedagogical agents into the design in order to make the tutorials more effective.

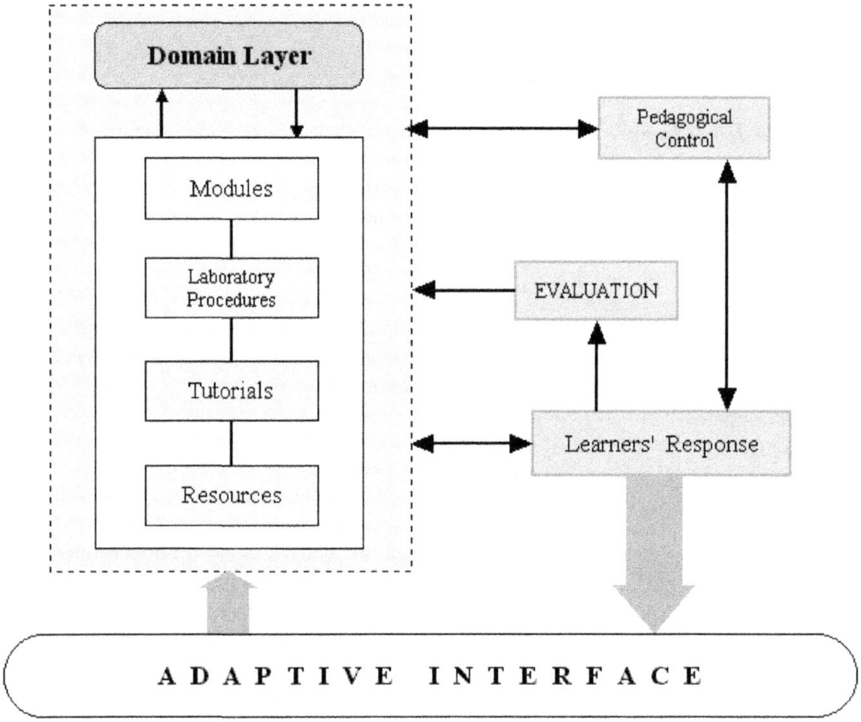

Fig. 1. System architecture of a pedagogical ITS module (Patil, 2004)

The system architecture used in the Web-Based Computer-Assisted Tutorials and Laboratory Procedures (WCALP) is based on *client-server* configuration. In this design, the user or client communicates with the server using a Web browser. This architecture has four important components, as outlined in the Figure 2:

- Server-side configuration;
- Library of software and database;
- Client-side configuration;
- User interface.

The user can access WCALP using any browser from his/her machine and perform the tutorial task including laboratory procedures. The responses from the user are returned to the WCALP server for further processing. The results will be calculated from the data collected and will then be analysed and sent back to the user's machine in order to check his/her performance. It is essential to provide the learners in any Web-based ITS with proper motivation for learning, the idea to recall the previous concepts learned, direction for essential immediate feedback and testing the performance. In order to achieve better learning outcomes, the sequence of the learning material can be managed and executed as per the hierarchy given by the flowchart developed.

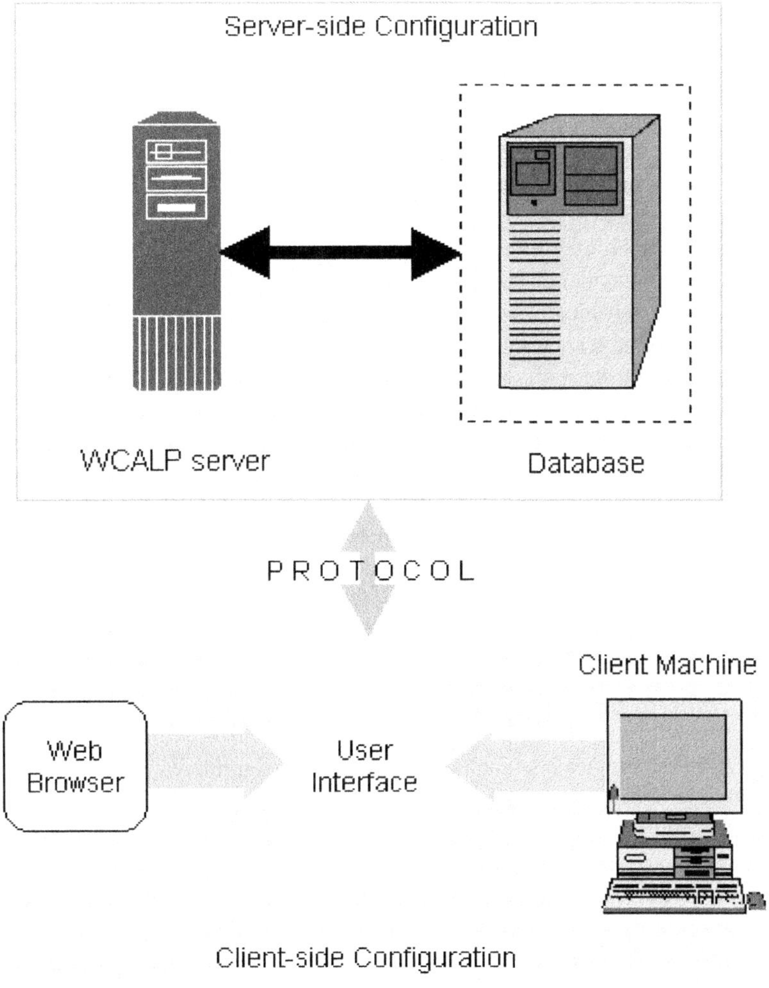

Fig. 2. System configuration of WCALP (Patil, 2004)

5.3 Authoring Software

Considering the applicability of the specialised programming language, such as Java (which contains multi-level function performing tasks), it is used as an authoring language to develop and design this project. Java is a fully object-oriented programming language and one of the fastest growing programming technologies of all time. Java represents an evolutionary change from earlier programming languages, rather than a radical departure. It is possible to make a Web page come alive with animations and user interactivity with Java. Java is the most important language for Web and Internet applications.

The WCALP consists of Java applets and Java script programs that are incorporated in the main page of each unit. This provides several exercises that are related to that unit and is useful in order to refresh and recall the necessary information of that particular unit. This type of design basically works with Transmission Control Protocol/Internet Protocol (TCP/IP). All Java applets are activated and opened on the user's (student's) machine with the help of the Java Virtual Machine (JVM) in the Web browser, which actually downloads the programming code and executes it there. As a result, a pop-up window containing an applet is opened that directly communicates with the server.

Although Java applets can provide strong client-side control, there are concerns related to the download and storage of Java applets in case when the file size is comparatively larger. Furthermore, applet functionality can vary between different browsers (Berntzen et al, 2001). The basic problems identified by Junglas (2003) with the developments of Java applets concern a huge amount of work involved and the requirement of profound Java programming knowledge. Java applets are truly platform independent and have in-built security against corrupt or bad coding of the system. However, running multiple and large files of Java applets on a small system can cause loading problems. In order to minimise the difficulties associated with applet configurations, only one or two applets are developed for each Unit, having a small file size that can be loaded easily and run on any system. The most relevant alternative to applets is to use JavaScript.

5.4 Immediate Feedback

The tutorial strategies also include the provision of immediate feedback. The central idea of learner interface employed in this project is to record the learner's response into the server database, to provide immediate access to the correct answers or hints and also to provide detailed information about the learner's performance whenever required. This has been achieved by using simple scripting in the Web pages. For example, the *check answer* button in every question window has a link to the correct answer, which allows the user to check the correct answer with a pop-up answer window of that question. The simple script used for this is given in Appendix V.

5.5 *CGI Programming and Perl*

In the case of the WCALP, it is proposed to use an *MS Access* database where all question and answer files will be stored at the WCALP server. Using query generated by Perl scripting language, the questions can be accessed from the server and responses from the user can be stored back where they are processed and again sent back to the user to check on his/her performance. Although, CGI applications have a better security level, speed is the main concern associated with the CGI programming in the case of a large database, which has to be processed in order to give updated results to the user. Also, CGI scripts can only handle small applications.

6 Issues in Developing ITS and Intelligent CAI

Web-based education is being considered as the replacement for traditional classroom-based education at the tertiary level. Barretto et al (2003) claim that user activity and graphic design are two major attributes that influence the development of educational software projects. Although Web-based instructions have distinct advantages over traditional classroom-based learning, the design and development of appropriate Web-based instructional software is a great challenge. CAL or Web-based learning tools offer more dynamic representations of learning materials than traditional blackboard teaching. Although developing any Web-based educational system involves a series of challenges and issues, three major factors of the development that must be considered are namely:

- The development of instructional software;
- The development of the database management;
- The development of assessment tools.

The literature research on ITS shows that ITSs have not been completely successful when implemented on a larger scale, due to various problems associated with their developments. McArthur et al (1993) pointed out that ITS limitations are influenced by certain components, such as the expert system, pedagogical component, student model and interface. Kinshuk & Patel (1997) claim that the development of ITSs has suffered due to the lack of software development that can be specifically used for tutoring purposes. The important issues incurred during the development of WCALP are discussed below.

6.1 *Pedagogical Concerns*

ITS has many advantages given the recent advancements in systems of education, including online learning. However, there are few minor concerns and issues related to the development of ITS or Intelligent Computer-Assisted Instruction. Most of ITS developers lack understanding of the implementation of pedagogical concepts that need to be incorporated within the ITS; consequently, the end-product design may not fulfil all of the pedagogical elements required in the tutoring strategies.

6.2 Software Latency

Hardware and software or programming languages are other important concerns associated with AI-based ITS. There are several advanced computer programming languages currently available, yet the application of proper computer language is one of the essential needs in developing intelligent tutoring. Kinshuk & Patel (1997) claim that the development of ITSs has suffered due to the lack of development of underlying software engineering methodologies for tutoring purposes. This gives rise to a need for generic software that can be rapidly customised as per individual needs. Also, the important concern is the design and maintenance of the large database used in these systems. Since ITS or AI-based tutoring consist of several files of information in the form of text, graphics and audio-visual data, it is somewhat cumbersome to manage all of this data. Also, the research shows that in some of the ITSs, debugging is the main concern in the designed software; it is very difficult to debug problems encountered in authoring programs due to large file sizes and complicated interfaces.

6.3 Courseware Development

Any computer-assisted learning or ITS must have essential courseware design that can easily fulfil the required pedagogical requirements in the learning strategy. In designing the project pedagogy for WCALP, several important factors have been taken into consideration, such as: the content of the course, learner's activities, performance indicators and learning outcomes. The WCALP has been developed using a framework in which three key strategies have been embedded in the model, namely: the tutorial strategy, hint strategy and evaluation strategy. This framework is very much pertinent in the development of Web-based educational software to help teaching/learning.

6.4 Authoring Software

The instructional software designed has the following important characteristics that are essential in order to implement the developed system with the appropriate instructional strategies. These characteristics are pertinent in the design of ideal instructional software for Web-based education in the engineering domain. These characteristics are listed as follows:

- The software is user friendly, easy to use and task-oriented.
- The instructional design strategies used in the software are able to provide the most appropriate encouragement for self-motivation.
- The software incorporates simple Web delivery methods with the minimum standards required of the desired content.
- The developed software provides realistic solutions to the problems according to the complex to simple principle.

- The software is learner-centred and not developer-centred, which provides opportunities for learners to test theories and explore their own learning.
- The software also develops active learning and facilitates essential thought processes, which result in a positive impact on the learning outcomes.
- The software utilises appropriate technology and includes educational facilities for self-assessment by learners throughout the learning process (Patil, 2004).

7 Conclusions and Future Work

The Web-based Intelligent Tutoring Systems (ITSs) are developing rapidly since the inception of the World Wide Web; however, very few of these are dedicatedly designed and developed for engineering domain so far. In this chapter, we described the methodology on the effectiveness of the developed Web-based intelligent and interactive systems in order to test its potential in learning pedagogy. The novel learning pedagogy applied in the circuit theory for basic electrical engineering has been also explained with the developed interactive Web pages. The next major goal for this developed system is to research on its effectiveness by data collection and analysis. This research can also be extended to develop and evaluate the Web-based model for other engineering related domains in the future.

7.1 Software Updating

The language used in the development of WCALP is Java in order to incorporate interactive tutorial components. Java programming has the advantage of an object oriented programming facility that supports several types of Web developments. However, the use of Java applets in Web applications has been limited due to its downloading problems and functionality with different browsers. As a result, it is envisaged that either advanced JSP or Java Servlet technology.

7.2 Using Advanced Java Solutions

With regard to JSP or Java Servlets, a dedicated server will be used, such as Apache Web server or Apache JServ servlet server. This type of server configuration and a combination of Java with other application-programming interface can provide efficient multi-user, real time interaction (Chang & Hung, 2000). This will help in enabling the use of several Web server systems in various formats like HTML, JavaScripts etc.

7.3 Using Commercial Software

Recent developments in computer-assisted instruction have shown that several interactive authoring solutions for computer-delivered control, simulation and scientific visualisations are available in the market (Pudlowski, 1995b). Evidence has also been found of similar developments that have been generated with the aid of several types of commercial authoring software, such as *Authorware*, *Director*, *LabVIEW*, *Hyeprtext*, *WebCT*, etc. Recent developments of LabVIEW-based applications have shown that many classroom-based engineering courses and conventional laboratory work can be combined with custom-written virtual instrumentation by computer-assisted delivery (Ertugrul, 2000).

Another option in the future developments of the WCALP is to utilise and test the *LabVIEW* package for advanced laboratory simulations and experiments in circuit theory. Since several commercial authoring software packages available are very complex to use, expensive and lack all of the functions required for certain learning tasks (Ertugrul, 2000), the possible integration of the *LabVIEW* package will be tested as pilot project for its ease of use and effectiveness. The advantage of using these types of readymade software tools is that they provide flexible navigation and other designed tools so that developers can concentrate more on pedagogical design issues, rather than technological aspects.

7.4 Adding Interactive Hypermedia and Animations

In future developments, interactive hypermedia and animations may be added to the WCALP with the integration of Flash technology. In using this combination of Java and Flash, learners can easily interact with the animations and demonstrations in learning theories. Computer-based simulations provide an excellent environment as an advanced teaching tool for several subjects in engineering and technology education (Roorda & Cartwright, 1994, Kanflo & Sagert, 1994, Veith et al, 1998). When applied to engineering education, the benefits of Web-based simulations as pointed out by Veith et al (1998), include controlled access, wide accessibility, increased integration and efficient maintenance. This is possible with the integration of the object oriented programming language like Java. According to Veith et al (1998), Internet-based simulations can help educators incorporate simulation models into their courses without cost, as well as avoiding inconveniences that arise due to the special equipment requirements.

References

1. Aberek, B., Popov, V.: Intelligent tutoring system for training in design and manufacturing. Advances in Engineering Software 35(7), 461–471 (2004)
2. Albacete, P.L., VanLehn, K.: The conceptual helper: an intelligent tutoring system for teaching fundamental physics concepts. In: Gauthier, G., VanLehn, K., Frasson, C. (eds.) ITS 2000. LNCS, vol. 1839, pp. 564–573. Springer, Heidelberg (2000)
3. Atkinson, R.C., Wilson, H.A. (eds.): Computer-Assisted Instructions. Academic Press, New York (1969)
4. Barretto, S.F.A., Piazzalunga, R., Ribeiro, V.G., Dalla, M.B.C., Filho, R.M.L.: Combining interactivity and improvement layout while creating educational software for the Web. Computers and Education 40(3), 271–284 (2003)

5. Berntzen, R., Strandman, J.O., Fjeldly, T.A., Shur, M.S.: Advanced solutions for performing real experiments over the Internet. In: International Conference on Engineering Education, Oslo, Norway (2001), `http://www.ineer.org/Events/ICEE2001/Proceedings/Index.htm` (Retrieved November 18, 2003)
6. Boulay, B.d.: What does the "AI" in AIED buy? In: Colloquium on Artificial Intelligence in Educational Software, IEE Digest No: 98/313, pp. 3/1–3/4 (1998)
7. Chang, T., Hung, D.: Web-based distance experiments: design and implementation. In: IEEE International Conference on Engineering Education, Taipei, Taiwan (2000), `http://www.ineer.org/Events/ICEE2000/Proceedings/papers/WA4-4.pdf`
8. Crowley, R.S., Medvedeva, O.: An intelligent tutoring system for visual classification problem solving. Artificial Intelligence in Medicine 36(1), 85–117 (2006)
9. Curilem, S.G., Barbosa, A.R., de Azevedo, F.M.: Intelligent tutoring systems: Formalization as automata and interface design using neural networks. In: Computers and Education (2006) (in press), `http://dx.doi.org/10.1016/j.compedu.2005.10.005`
10. Dear, B.L.: Artificial Intelligence Techniques: Applications for Courseware Development. In: Expert Systems and Intelligent Computer-Aided Instructions, pp. 91–99. Educational Technology Publications, Englewood Cliffs (1991)
11. Ertugrul, N.: Towards virtual laboratories: a survey of *LabVIEW*-based teaching/learning tools and future trends. International Journal of Engineering Education 16 (3), 171–180 (2000)
12. Fang, Y., Liou, F.W.: Computer simulation of three-dimensional mechanical assemblies: part-i – general formulation. In: 1993 ASME Inter. Computers in Education Conference and Exhibition, pp. 579–587 (1993)
13. Godwin-Jones, B.: Emerging technologies, dynamic Web page creation. Language, Learning and Technology Journal 1(2), 7–13 (1998), `http://polyglot.cal.msu.edu/llt/vol1num2/emerging/default.htm` (Retrieved March 12, 2002)
14. Graham, D.T., McNeil, J., Pettiford, L.: Untangled Web: Developing Teaching on the Internet. Pearson Education, Harlow (2000)
15. Gwo-Jen Hwang, G.-J.: A conceptual map model for developing intelligent tutoring systems. Computers and Education 40(3), 217–235 (2003)
16. Han, S.-G., Lee, S.-G., Jo, G.-S.: Case-based tutoring systems for procedural problem solving on the www. Expert Systems with Applications 29(3), 73–582 (2005)
17. Hativa, N.: Students' conceptions of, and attitudes toward, specific features of a CAI system. Journal of Computer-Based Instruction 16(3), 81–89 (1989)
18. Huang, C.-J., Chu, S.-S., Guan, C.-T.: Implementation and performance evaluation of parameter improvement mechanisms for intelligent e-learning systems. Computers and Education (2006) (in press), `http://dx.doi.org/10.1016/j.compedu.2005.11.008`
19. Jaques, P.A., Vicari, R.M.: A BDI approach to infer student's emotions in an intelligent learning environment. Computers and Education (2006) (in press), `http://dx.doi.org/10.1016/j.compedu.2005.09.002`
20. Junglas, P.: Using applets for physics education: a case study of non-linear systems and chaos. In: 7th Baltic Region Seminar on Engineering. Education, pp. 61–64 (2003)
21. Kanflo, T.J., Sagert, P.A.: Teaching the design of manufacturing systems using simulation. European Journal of Engineering Education 19(1), 117–123 (1994)

22. Kinshuk, Patel, A.: A Conceptual Framework for Internet Based Intelligent Tutoring Systems. In: Behrooz, A. (ed.) Knowledge Transfer, vol. 2, pp. 117–124. pAce, London (1997)
23. Kulik, C.-L., Kulik, J.A.: Effectiveness of computer-based instruction: an updated analysis. Computers in Human Behaviour 7(1-2), 75–94 (1991)
24. Larkin, J.H., Chabay, R.W. (eds.): Computer-Assisted Instruction and Intelligent Tutoring Systems: Shared Goals and Complementary Approaches, pp. 151–186. Lawrence Erlbaum & Assoc., Hillsdale (1992)
25. Murray, T.: Expanding the knowledge acquisition bottleneck for intelligent tutoring systems. International Journal of Artificial Intelligence in Education 8(3) (1997)
26. Olsen, J.B.: The Four Generations of Computerised Testing: toward Increased Use of AI and Expert Systems. In: Expert Systems and Intelligent Computer-Aided Instructions, pp. 66–71. Educational Technology Publications, Englewood Cliffs (1991)
27. Ong, J., Ramachandran, S.: Intelligent tutoring systems: the what and the how. Learning Circuits (2000), `http://www.learningcircuits.org/feb2000/` (Retrieved June 15, 2003)
28. Patil, A.S.: Issues in the design and development of Web-based computer-assisted tutorials and laboratory procedures in basic electrical engineering. Unpublished masters dissertation, Monash University, Melbourne (2004)
29. Pudlowski, Z.J.: Major Issues in Developing Computer-Aided Authoring Programs for Engineering Education. In: Pudlowski, Z.J., et al. (eds.) Computers in Electrical Engineering Education – Research, Development and Application, pp. 23–31. EEERG & USICEE, Melbourne (1995)
30. Reed, J.A., Afjeh, A.A.: Developing interactive educational software for the World Wide Web with Java. Computers and Education 30(3-4), 183–194 (1998)
31. Roorda, J., Cartwright, T.: Engineering design simulation: a computer-based teaching environment. European Journal of Engineering Education 19(1), 105–115 (1994)
32. Schacter, J.: The Impact of Education Technology on Student Achievement: What the Most Current Research has to Say (2003), `http://www.mff.org/publications/publications.taf?page=161` (Retrieved October 3, 2003)
33. Sleeman, D., Brown, J.S.: Intelligent Tutoring Systems. Academic Press, London (1982)
34. Stathacopoulou, R., Magoulas, G.D., Grigoriadou, M., Samarakou, M.: Neuro-fuzzy knowledge processing in intelligent learning environments for improved student diagnosis. Information Sciences 170(2-4, 25), 273–307 (2005)
35. Surber, J., Leeder, J.: The effect of graphic feedback on students' motivation. Journal of Computer-Based Education 15(1), 14–17 (1988)
36. Thomas, E.: Intelligent Tutoring Systems, ITS (2003), `http://coe.sdsu.edu/eet/Articles/tutoringsystem/start.htm` (Retrieved June 15, 2003)
37. Toogood, R.W., Wong, D.: An interactive fluid mechanics tutorial using authorware. In: 1993 ASME International Computers in Education Conference and Exhibition, pp. 539–545 (1993)
38. Veith, T.L., Kobza, J.E., Koelling, C.P.: World Wide Web-based simulations. International Journal of Engineering Education 14(5), 316–321 (1998)
39. Wankat, P.C., Oreovicz, F.S.: Teaching Engineering. McGraw-Hill, New York (1993)

40. Wedekind, J.: Computer-Aided Model Building. In: Mandl, H., Lesgold, A. (eds.) Learning Issues for Intelligent Tutoring Systems, pp. 287–294. Springer, New York (1988)
41. Wilson, B.G., Welsh, J.R.: Small Knowledge-Based Systems in Education and Training: Something New under the Sun. In: Expert Systems and Intelligent Computer-Aided Instruction, vol. 2. Educational Technology Publications, Englewood Cliffs (1991)
42. Woo, C.-W., Evens, M.W., Freedman, R., Glass, M., Shim, L.-S., Zhang, Y., Zhou, Y., Michael, J.: An intelligent tutoring system that generates a natural language dialogue using dynamic multi-level planning. In: Artificial Intelligence in Medicine (2006) (in press), http://dx.doi.org/10.1016/j.artmed.2005.10.004
43. Woolf, B.P.: Building knowledge based tutors. In: Tomek, I. (ed.) ICCAL 1992. LNCS, vol. 602, pp. 46–60. Springer, Heidelberg (1992)
44. Yaakub, M.N.: Effectiveness of Computer-Assisted Instruction in Technical Education: a Meta-Analysis (2001), http://voc.ed.psu.edu/projects/publications/books/Fall%202001/WEF2001.2.html (Retrieved May 12, 2003)
45. Yang, H., Lu, W.F.: Procase: a prototype of intelligent case-based process planning system with simulation environment. In: 1993 ASME International Computers in Education Conference and Exhibition, pp. 571–577 (1993)

Granular Mining of Student's Learning Behavior in Learning Management System Using Rough Set Technique

Nor Bahiah Hj Ahmad[1], Siti Mariyam Shamsuddin[1], and Ajith Abraham[2]

[1] Faculty of Computer Science and Information Systems,
Universiti Teknologi Malaysia,
81310 Skudai, Johor, Malaysia
bahiah@utm.my, mariyam@utm.my
[2] Centre for Quantifiable Quality of Service in Communication Systems,
Norwegian University of Science and Technology, Trondheim, Norway
ajith.abraham@ieee.org

Abstract. Pattern multiplicity of user interaction in learning management system can be intelligently examined to diagnose students' learning style. Such patterns include the way the user navigate, the choice of the link provided in the system, the preferences of type of learning material, and the usage of the tool provided in the system. In this study, we propose mapping development of student characteristics into Integrated Felder Silverman (IFS) learning style dimensions. Four learning dimensions in Felder Silverman model are incorporated to map the student characteristics into sixteen learning styles. Subsequently, by employing rough set technique, twenty attributes have been selected for mapping principle. However, rough set generates a large number of rules that might have redundancy and irrelevant. Hence, in this study, we assess and mining the most significant IFS rules for user behavior by filtering these irrelevant rules. The assessments of the rules are executed by evaluating the rules support, the rules length and the accuracy. The irrelevant rules are further filtered by measuring different rules support, rules length and rules accuracy. It is scrutinized that the rules with the length in between [4,8], and the rules support in the range of [6,43] succumb the highest accuracy with 96.62%.

Keywords: Rough set, learning styles, discretization, rule generation, rule filtering, classification.

1 Introduction

Learning Management System (LMS) is an environment to support web-based learning content development. The main features of an LMS include content creation, content repository management, content delivery and interface, and learning process management such as course enrollment, assessment and performance tracking [1]. Educators are able to distribute information to students, produce content material, prepare assignments and tests, engage in discussions, manage

F. Xhafa et al. (Eds.): Computational Intelligence for Tech. Enhanced Learning, SCI 273, pp. 99–124.
springerlink.com

distance learning and enable collaborative learning using forums, chats and news services. Several examples of popular LMS are Moodle [2], Blackboard [3] and WebCT [4]. Recently, Moodle, an acronym for Modular Object-oriented Dynamic Learning Environment has become one of the most commonly used LMS.

Moodle is an open source LMS that enable the creation of powerful, flexible and engaging online courses and experiences. Several e-learning researches have been conducted in order to enhance Moodle's performance [1, 5, 6]. Graf and Kinshuk [5] has extended Moodle's capability by implementing adaptation of the learning material based on the student's learning style. A standalone tool for automatic detection of learning styles in LMS has been implemented. Kerdprasop et al. [1] enhanced LMS functionality to individualize the learning content with induction ability. Romero et al. [6] had developed data mining tool to help instructors preprocess or apply mining techniques, such as statistics, visualization, classification, clustering and association rule mining from Moodle data. E-learning systems developed using Moodle accumulate an enormous amount of information which is very valuable for analyzing students' behavior and could create a gold mine of educational data [7]. Any LMS data can be mined in order to understand the dynamic behavior of students in the web. Such information can be used to improve the implementation of e-learning system by determine e-learning effectiveness and measure the success of various instructional effort [8].

Rough set theory, introduced by Zdzislaw Pawlak in the early 1980's is a mathematical tool to deal with vagueness and uncertainty [37]. The methodology is concerned with the classificatory analysis of vague, uncertain or incomplete information or knowledge expressed in terms of data acquired from experience. Unlike other soft computing methods, rough set analysis requires no external parameters and uses only the information presented in the given data. Rough set method does not need membership functions and prior parameter settings. It can extract knowledge from the data itself by means of indiscernibility relations and generally needs fewer calculations compare to fuzzy set theory. The attribute reduction algorithm removes redundant information or features and selects a feature subset that has the same discernibility as the original set of features. The selected features can describe the decision as well as the original whole features set, leading to better classification accuracy.

Meanwhile, the flexibility offered in any LMS allow the instructor to design and deliver various sources of learning material such as animation, power point, video, hypermedia and on-line tutorial easily. However, each individual student has their own learning preferences in order to comprehend the knowledge. They learn better when they are given a learning environment that is suitable with their learning style. By employing rough set, it is crucial to select the most vital attributes influencing the learning behavior of a student. The chosen attributes, observed from student behavior in Moodle are then engaged within a decision rule generation process, creating descriptive rules for the classification task. Decision rules extracted by rough set algorithms are concise and valuable, which can be benefit to identify student's learning style by enlightening some knowledge hidden in the data.

This chapter provides in-depth discussions on student's learning preferences and behavior while using e-learning system based on Felder Silverman learning dimension, such as processing, perception, understanding and input. The student's attributes are fed to rough set classifier to obtain the granularity of the significant features. To our understanding, none of the studies have been reported in implementing IFS features with significant rules for identifying learning styles. However, our focus will be more on extracting the significant rules for detecting the learning styles which is not previously done by other researchers in this area.

This chapter is organized as follows: Section 2 explains previous studies and issues related to student's behaviour while learning on-line, and followed by section 3 that describes the proposed integrated Felder Silverman learning style model. The development of e-learning system that incorporates learning resources for Felder Silverman learning dimension is described in section 4. Section 5 gives in-depth analysis on students' behavior while learning using hypermedia learning system. These include the analysis of the learners learning style distributions, preferences and their navigation behavior. The analysis is useful for providing parameters for classification of student's learning style based on student's learning characteristics while learning online.

In section 6, we present an intelligent data analysis approach based on rough set for generating classification rules from a set of data samples describing the student's behavior and activities during e-learning session extracted from Moodle log files. Section 7 discusses rules filtering approach to identify the most significant rules. Based on the generated rules, the diagnosis is executed to map student's learning style into IFS features. Section 8 gives conclusion of the chapter.

2 Related Work

The early research that concern with student's learning style used questionnaire to assess the student's learning characteristics [9, 10, 11, 12]. However, the exploitation of questionnaires is time consuming and unreliable approach for acquiring learning style characteristics and may not be accurate [13, 14, 15]. Most questionnaires are too long, hence, causing students to choose answers arbitrarily instead of thinking seriously about them. Even if the learning style has been determined, it still cannot notify the real characteristics of the students while learning on-line. In addition, once the profile is generated, it becomes static and doesn't change regardless of user interaction. In on-line learning environment, the student's learning characteristics are changed accordingly when different tasks are provided. Due to these problems, several studies have been conducted in detecting student's learning style that are based on the student's browsing behavior [16, 17, 18, 19]. This approach can be implemented successfully since the style of student's interaction with the system can be inferred accurately and can be used as attributes for adaptation purposes.

Various techniques have been used to represent student learning style such as statistics [17], Neural Network [19, 13], Decision Tree [20], Bayesian Networks [16], Naïve Bayes [15] and Genetic Algorithm [18]. Previously, we have successfully classified students' learning style using Backpropagation Neural Network (BPNN) [21]. However, BPNN lacks of explanatory power and difficult to identify rules concerning to the inputs and outputs [22].

Table 1 list several studies focusing on learning style detection for the past five years. Various techniques, approaches and purposes of detecting learning styles have been discussed extensively in the literature. It can be observed from the pattern of research conducted, the trend of research on detecting student's behavior recently have focus more on LMS. The trend shows that LMS has become a popular tool in developing e-learning materials due to the flexibility and robustness provided by the tool.

Table 1. Related studies on learning style detection

Author	Learning Style Model	E-learning data	Identification Technique
Kelly and Tangney (2004) [15]	Garderners Multiple Intelligence	On-line learning system	Naïve Bayes
Lo and Shu (2005) [19]	Kinestethic, Visual, Auditory	On-line learning system	Neural Network
Garcia P. et al (2007) [16]	Felder Silverman – 3 Dimension	On-line learning system	Bayesian Network
Villaverde J.E. et al. (2006) [13]	Recognition of learning style in e-learning – Felder Silverman	Simulated Data	Neural Network
Cha H. et al. (2006) [20]	Felder Silverman	On-line learning system	Decision Tree
West et al (2006) [23]	Investigate impact of learning style on e-learning	On-line class	Statistics
Ai and Laffey (2007) [8]	Pattern classification of student's performance	LMS	Web Mining
Garcia E. et al. (2007) [24]	Student activities and behavior	LMS	Association rule mining
Kerdprasop et al (2008) [1]	Classify student knowledge level	LMS	Rule induction rough set
Graf S. and Kinshuk (2008) [25]	Analysing Student's Behavior based on Felder Silverman	LMS	t-test, u-test
Romero et al (2008) [6]	Develop mining tool	LMS	Data mining

On the other hand, the common e-learning system, Moodle, was developed based on the social constructivism pedagogy support communication and collaboration among communities of learning through the activity modules which includes Forums, Wikis, Databases, Discussion Board, Chats, and Journals. Student can actively participate in discussion forum to discuss certain subject matter or to solve certain task assigned to them either individually or in a team. Moodle also supports import and export of SCORM/IMS Content Packaging standards. Hence, educators are able to create learning objects to deliver learning content and exercises to students and assess learning using assignments or quizzes. The options for importing learning objects from other sources enable educators to create very rich learning resources. Student activities, such as reading writing, submitting assignments, taking tests, performing various tasks and even communicating with peers were recorded and kept in log files [7]. User's personal information, their academic results and their history of interaction were kept in data base and very useful for educators to enhance learning environment.

Based on the flexibility offered in Moodle, this study has developed the learning resources for Data Structure Subject that consider Felder Silverman Learning Style model. The student activities while using the learning resources is analyzed in order to determine the relationship between the students's learning style and their learning preferences. In this chapter we identify significant parameters for classifying students learning style based on their characteristics while learning online. The information is needed to be stored in user profile for adaptation of learning materials based on learning style. In our study, we will classify the proposed IFS features of students learning styles by employing rough set technique granularity mining. This includes extracting the significant rules for detecting the learning styles.

3 The Proposed Integrated Felder Silverman Learning Style Model

A variety of learning style model has been used to characterize learning styles for students. Among them are Felder Silverman learning style [26], Kolb's theory of experiential learning [27], Howard Gardner Multiple Intelligence [28], Honey and Mumford [29] and Dunn and Dunn model [30]. In this work, we utilize Felder Silverman model to investigate the student's preferences of the learning material. The reason behind choosing this model is due to a considerable amount of literature on this subject.

The model has been successfully used in previous studies involving adaptation of learning material, collaborative learning and for traditional teaching [26, 31, 32]. Furthermore, the development of the hypermedia learning system that incorporate learning components such as the navigation tool, the presentation of the learning material in graphics form, simulation, video, sound and help facilities can easily tailored to the Felder Silverman learning style dimension.

Felder Silverman learning style model was developed by Felder and Silverman in 1998 [33]. This model categorized a student's dominant learning style along a scale of four dimensions: active-reflective (how information is processed), sensing-intuitive (how information is perceived), visual-verbal (how information is presented) and global–sequential (how information is understood). Table 2 describes the characteristics of students based on the learning dimensions.

Felder and Solomon developed Index of Learning Styles (ILS) questionnaire to identify the student's learning style. The objective of this questionnaire is to determine the dominant learning style of a student. This questionnaire can be accessed freely from web site [34] and is often used as instrument to identify learning style. The ILS questionnaire consists of 44 questions with two possible answers, A or B. These questions are separated into four groups, with eleven questions each. These groups correspond to four categories of Felder Silverman learning dimension (active-reflective, sensing-intuitive, visual-verbal, and sequential-global).

Table 2. Felder Silverman learning dimension and learner characteristics [33]

Learning Dimension	**Learner Characteristics**	
	Active	**Reflective**
Processing	Retain and understand information best by doing something active with it such as discussing it, applying it, or explaining it to others.	Prefer observation rather than active experimentation. Tend to think about information quietly first.
	Sensor	**Intuitive**
Perception	Like learning facts, often like solving problems by well-established methods and dislike complications and surprises. Patient with details and good at memorizing facts and doing hands-on work. More practical and careful than intuitors.	Prefer discovering possibilities and relationships. Like innovation and dislike repetition. Better at grasping new concepts and comfortable with abstractions and mathematical formulations. Tend to work faster and more innovative than sensors.
	Visual	**Verbal**
Input	Remember best what they see from visual representations such as graphs, chart, pictures and diagrams.	More comfortable with verbal information such as written texts or lectures.
	Sequential	**Global**
Understanding	Prefer to access well structured information sequentially, studying each subject step by step.	Prefer to learn in large chunks, absorb material randomly without seeing connections and then suddenly getting it. Able to solve complex problems quickly or put things together in novel ways once they have grasped the big picture.

Fig. 1 shows the learning style scales for each learning style dimension. The score are expressed with values between 11A to 11B for each dimension. If a student gets a score from 1 to 3 in any dimension, he/she has a mild preference and fairly balanced on the two dimensions. If the score is on scale 5 or 7, the student has moderate preference, and if the score is on scale 9 or 11 the student has a very strong preference for the dimension. The student with strong preferences for certain dimension must learn according to the environment that matches his learning style. He may have learning difficulty if he studies in the environments that are not suitable with his learning style.

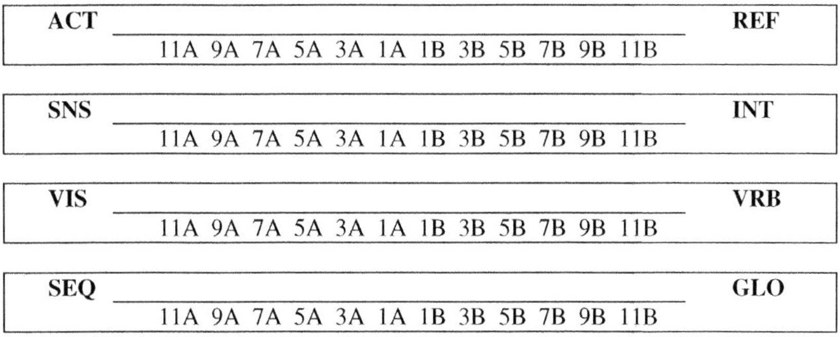

Fig. 1. Felder Silverman learning style scales [34]

Table 3. 16 learning styles based Integrated Felder Silverman learning dimension

Learning Styles	Label
Active/Sensor/Visual/Sequential	ASViSq
Reflective/Sensor/Visual/Sequential	RSViSq
Active/Intuitive/Visual/Sequential	AIViSq
Reflective/Intuitive/Visual/Sequential	RIViSq
Active/Sensor/Verbal/Sequential	ASVbSq
Reflective/Sensor/Verbal/Sequential	RSVbSq
Active/Intuitive/Verbal/Sequential	AIVbSq
Reflective/Intuitive/Verbal/Sequential	RIVbSq
Active/Sensor/Visual/Global	ASViG
Reflective/Sensor/Visual/Global	RSViG
Active/Intuitive/Visual/Global	AIViG
Reflective/Intuitive/Visual/Global	RIViG
Active/Sensor/Verbal/Global	ASVbG
Reflective/Sensor/ Verbal/Global	RSVbG
Active/Intuitive/Verbal/Global	AIVbG
Reflective/Intuitive/Verbal/Global	RIVbG

Without considering the scale, this study integrate the processing, perception, input and understanding learning styles to map the characteristics of the students into 16 learning style. Table 3 depicts the 16 learning styles that are proposed in Integrated Felder Silverman model. The rationale of the integration of these dimensions is to minimize the time consumption in diagnosing the learning styles.

4 Development of the Learning System

When designing the learning material, it is important to accommodate elements that reflect individual differences in learning [35]. Systems such as iWeaver [9], INSPIRE [10], CS383 [11] and SAVER [19] proposed several learning components to be tailored according to learning styles of the students. We adopted the components implemented in iWeaver, CS383 and INSPIRE in our system due to the success of the researches. The resource materials are categorized into communication tool using forum facility, learning resources, exercises and examples.

By utililizing Moodle features, the learning resources developed for this study were structured into components that are suitable for processing, perception, input and understanding dimension in Felder Silverman learning style model. Three chapters from Data Structure syllabus such as, Sorting, Searching and Linked List were integrated in the system. Among the resource materials provided in the learning systems are as follows:

- **Forum** – Provide mechanism for active discussions among students. Based on certain topics assigned, student's can post, reply and view discussions conducted in the forum. The activity is very useful for active learners, while reflective learners usually prefer to view the content of the forum.
- **Animation** – Provide simulations of various sorting, searching and link list operations. The process of how each technique is implemented can be viewed step by step according to the algorithm. The activity is useful specifically for visual learners.
- **Sample Codes** – Provide source codes for every algorithms discussed in class. Students can view and actively run various programs on sorting, searching and link list. They also were given tasks to do some modification to the source codes.
- **Hypertext** – Provide learning content which has been composed into theory and concepts. The learning content has the topic objectives, sub modules, and navigation link. This learning resource is useful for sensor/intuitive students and sequential/global students. Sensor students can understand well fact materials, while intuitive learners can learn well the abstract materials.
- **Power Point Slideshow** – Provide learning materials that consist of example in the form of text, pictures and animations. Different colors of text are used to emphasized different facts in the example given.
- **On-line exercises** – Provide exercises in multiple choice questions which students can answer and get hint and feedback regarding their performance.

- **On-line Assessment**– Provide on-line quiz that consist of multiple choice questions and marks that can be displayed immediately after the student submit the quiz. Analysis of each item in the question is also provided in Moodle.

Fig. 2 is an example of the e-learning system developed for Data Structure subject. The system consists of the resource materials listed previously. The flexibility provided by Moodle allows the resource materials to be developed easily without requiring expertise in managing the e-learning system.

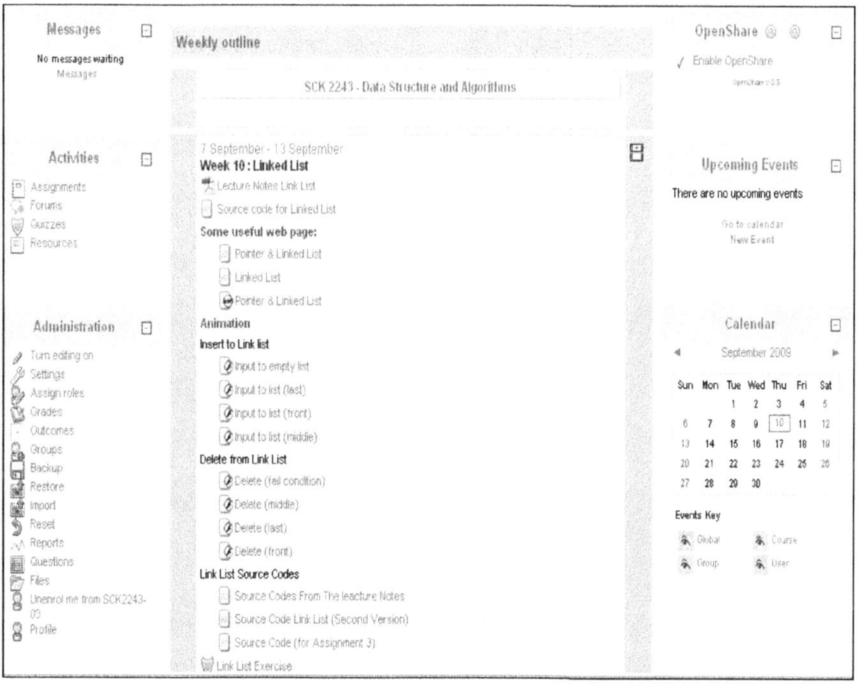

Fig. 2. Example of e-learning developed using Moodle

5 Experimental Setup

In order to determine which characteristics of the students can be used to identify their learning style, we have conducted experiments for 2 semesters. There were 110 students participated in this study. The experiments took place at Faculty of Computer Science and Information Systems, Universiti Teknologi Malaysia (UTM). Students' composition is categorized as follows: 70 students are from Computer Science stream and the other 40 students are from Computer Engineering stream. During study session, students' were required to participate in lab exercises, to work in group for problem solving, to explore self-study mechanism in using e-learning system, to take part in forum discussion and to acquire on-line quiz. The entire materials can be accessed through the e-learning system.

Fig. 3 shows the analysis process conducted in this study. Students were required to use the resource materials provided in the e-learning system. Every activity performed by the students, such as the student's involvement in forum, the frequency of accessing learning content, animation, exercises and on-line assessment were recorded and stored in log files. The data is pre-processed, analyzed and transformed into appropriate format in order to analyze and interpret the characteristics of the students based on Felder Silverman learning dimensions.

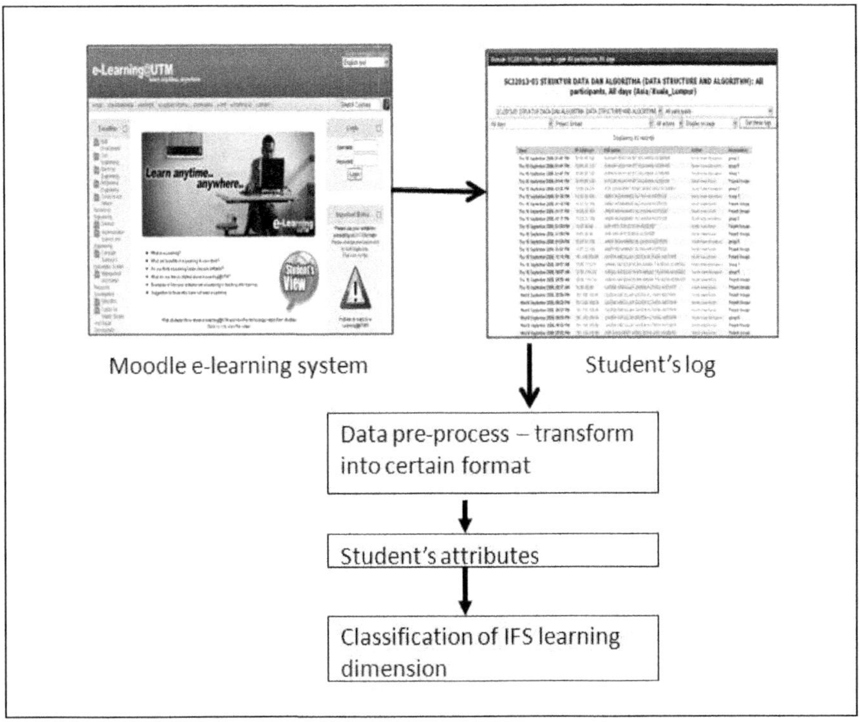

Fig. 3. The analysis process

5.1 Analysis of the Questionnaire

In order to determine the student's learning styles, the students were required to fill up the ILS questionnaire which can be accessed freely from website [34]. Fig. 4 shows the distribution of learning styles collected from 110 students taking Data Structure course in UTM. From the survey, we found out that there are only 13 learning styles exist among the group of students. Majority of the students, 28.2%, have Active/Sensor/Visual/Sequential learning styles followed by Reflective/Sensor/Visual/Sequential which is 19% among the sample group. The result is consistent with the studies done by Zywno [31] who concluded that the default learning styles among students is Active/Sensor/Visual/Sequential. However, in this study we found out that no students fall into 3 categories of learning style, which are Active/Intuitive/Verbal/Sequential, Reflective/Sensor/Verbal/Global and Reflective/Intuitive/Verbal/Global. The main reason for the absence is that there are only twelve students who have verbal learning styles. The small number of verbal students is not enough to cover the 16 learning styles.

The content of log files that contains every activity performed by the students, were transformed into appropriate format in order to analyze and interpret the characteristics of the students based on Felder Silverman learning dimensions. Fig. 5 shows an example of log data extracted by Moodle. The log contains the date and time of access, IP address, student name activity done during the interaction

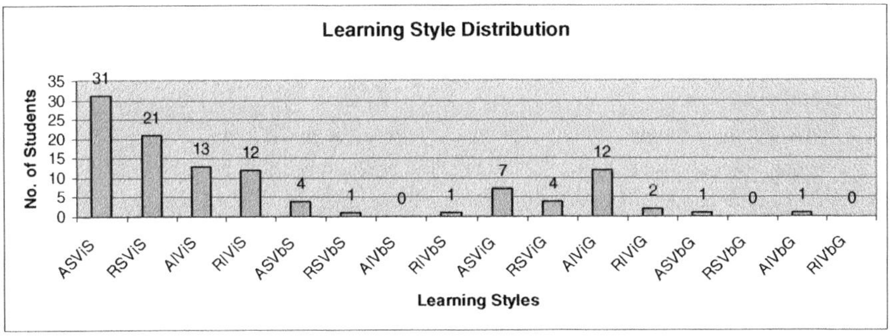

Fig. 4. Distribution of learning styles among UTM students

Home▶ SCJ2013-03▶ Reports▶ Logs▶ All participants, All days

SCJ2013-03 STRUKTUR DATA DAN ALGORITMA (DATA STRUCTURE AND ALGORITHM): All participants, All days (Asia/Kuala_Lumpur)

SCJ2013-03 STRUKTUR DATA DAN ALGORITMA (DATA STRUCTURE AND ALGORITHM) | All participants

All days | Project Groups | All actions | Display on page | Get these logs

Displaying 61 records

Time	IP Address	Full name	Action	Information
Thu 10 September 2009, 01:41 PM	10.60.85.120	NURAINI HIDAYAH BT SULAIMAN AC080405	forum view discussion	group 3
Thu 10 September 2009, 01:41 PM	10.60.85.120	NURAINI HIDAYAH BT SULAIMAN AC080405	forum view discussion	group 5
Thu 10 September 2009, 01:41 PM	10.60.85.120	NURAINI HIDAYAH BT SULAIMAN AC080405	forum view discussion	Group 1
Thu 10 September 2009, 01:41 PM	10.60.85.120	NURAINI HIDAYAH BT SULAIMAN AC080405	forum view forum	Project Groups
Thu 10 September 2009, 01:32 PM	10.60.84.224	NOR ASIAH BINTI MOHD MOHD ARIS AC080063	forum view discussion	group 3
Thu 10 September 2009, 01:19 PM	10.53.32.108	AMER MOHAMMED ALI HASAN AC070326	forum view discussion	Group 1
Thu 10 September 2009, 01:18 PM	10.53.32.108	AMER MOHAMMED ALI HASAN AC070326	forum view forum	Project Groups
Thu 10 September 2009, 01:17 PM	10.53.32.108	AMER MOHAMMED ALI HASAN AC070326	forum view forum	Project Groups
Thu 10 September 2009, 01:17 PM	10.53.32.108	AMER MOHAMMED ALI HASAN AC070326	forum add discussion	group 3
Thu 10 September 2009, 01:04 PM	10.60.90.49	NUR ANIS BTE NORDIN AC080067	forum view forum	Project Groups
Thu 10 September 2009, 01:04 PM	10.60.90.49	NUR ANIS BTE NORDIN AC080067	forum view forum	Project Groups
Thu 10 September 2009, 01:04 PM	10.53.32.108	AMER MOHAMMED ALI HASAN AC070326	forum view discussion	group 5
Thu 10 September 2009, 01:01 PM	10.53.32.108	AMER MOHAMMED ALI HASAN AC070326	forum view forum	Project Groups
Thu 10 September 2009, 12:15 PM	161.139.100.98	QASEM ABDULLAH QASEM ALTASHI AK070416	forum view forum	Project Groups
Thu 10 September 2009, 08:57 AM	10.60.114.210	AHMAD HAFIZUDDIN BIN AHMAD TAJUDDIN AC080383	forum view discussion	Group 1
Thu 10 September 2009, 08:57 AM	10.60.114.210	AHMAD HAFIZUDDIN BIN AHMAD TAJUDDIN AC080383	forum view discussion	group 5
Thu 10 September 2009, 08:56 AM	10.60.114.210	AHMAD HAFIZUDDIN BIN AHMAD TAJUDDIN AC080383	forum view forum	Project Groups
Thu 10 September 2009, 08:37 AM	10.60.85.88	QASEM ABDULLAH QASEM ALTASHI AK070416	forum view forum	Project Groups
Wed 9 September 2009, 06:59 PM	161.139.100.98	QASEM ABDULLAH QASEM ALTASHI AK070416	forum view forum	Project Groups
Wed 9 September 2009, 06:25 PM	161.139.100.98	QASEM ABDULLAH QASEM ALTASHI AK070416	forum view forum	Project Groups
Wed 9 September 2009, 06:07 PM	161.139.100.98	QASEM ABDULLAH QASEM ALTASHI AK070416	forum view forum	Project Groups
Wed 9 September 2009, 06:00 PM	161.139.100.98	QASEM ABDULLAH QASEM ALTASHI AK070416	forum view discussion	group 5
Wed 9 September 2009, 06:00 PM	161.139.100.98	QASEM ABDULLAH QASEM ALTASHI AK070416	forum view forum	Project Groups
Wed 9 September 2009, 05:53 PM	161.139.100.98	NOR ASIAH BINTI MOHD MOHD ARIS AC080063	forum view forum	Project Groups

Fig. 5. Sample log data in Moodle

and information of resources being accessed. Detailed discussions on the log analysis process can be referred in [36].

From the result of the analysis in [36], we conclude that the preferences of the students are consistent with the characteristics of the learning styles describe in Felder Silverman model. We further determine the parameters that represent the characteristics of the students based on Felder Silverman learning model. Table 4 lists the attributes for Felder Silverman learning dimension. The attributes can be used for mining the learning activities and preferences in e-learning environments. Such information is very useful for adaptation of learning materials and for analyzing the performance of the student's while learning on-line.

Table 4. Attributes description and values for IFS learning dimension

Attribute Name	Value	Dimension
Post and reply forum	Much/Few	
Number of exercise visited	Much/Few	
Number of simulation visited	Much/Few	Active/
Number of executing sample codes	Much/Few	Reflective
Number of viewing/reading forum content	Much/Few	
Hypertext coverage	Much/Few	
Number of backtrack in hypertext	Much/Few	
Viewing concrete material	Much/Few	
Viewing abstract material	Much/Few	Sensor/
Number of access to example	Much/Few	Intuitive
Number of exercise visited	Much/Few	
Exam delivery duration	Quick/slow	
Exam revision	Much/Few	
Number of Simulation visited	Much/Few	
Number of diagram/picture viewing	Much/Few	Visual/
Hypertext coverage	Much/Few	Verbal
PowerPoint Slide Access	Much/Few	
Hypertext – navigate linearly	Linear/global	
Hypertext coverage	Much/Few	Sequential
Number of visiting course overview	Much/Few	/global

6 Rough Set in Detecting Student Learning Style

Rough set offers some important techniques in managing an information system (IS), and consists of several steps leading towards the final goal of generating rules from information/decision systems. The main steps of the rough set approach are given below and the detail of each procedure can be found in [37].

- Mapping of information from the original database into the decision system format -Information System Table.
- Data Completion
- Data Discretization
- Reduct Computation
- Rules Derivation
- Rules Filtering
- Classification

Rough set approach is frequently used on attribute selection and feature selection [38, 39]. Redundant attributes will be removed to generate association rules that are more efficient. However, rough set usually generate an excessive amount of rules which might include redundant and duplicate rules. Therefore, it is important to mine the most significant rules to produce accurate classification since the main issue in rough set is to automatically extract an ideal, optimal and important rule. Variety of metrics has been used to measure rules generated by rough set, and these include rule confidence, accuracy, support, gain, chi-squared value, gini, entrophy gain and conviction. Many studies have been reported in mining the best rule; Li, 2006 [39] provide a rank of how important is each rule by rule importance measure, Ohrn, 1999 [40] used support/confidence value and [41, 42, 43] investigated the relationship of the rule support, length, and various number of rules at the classification phase. Bose, 2006 concluded that only the top 10% of the rule were really important and needed to be retained [41].

A general framework of rough set classification for detecting student learning style is presented in Fig. 6. The raw input data is designed based on the analysis of the student's learning style. The input data set is transformed into a decision system and pre-analysis is executed by eliminating data with incomplete value. The descretization process is employed for training and testing data set. The rules are

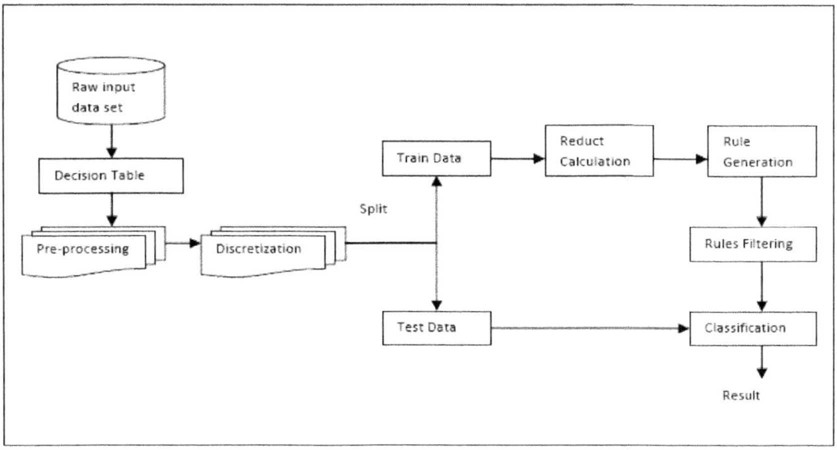

Fig. 6. Rough set frameworks for detecting students' learning style

generated from reducts by performing the attribute reduction algorithm. Consequently, these rules are filtered to get the most significant rule for the classification process. Detail discussions on classification and rule filtering process are given in the next subsection.

6.1 Data Design and Knowledge Representation

From the analysis of the learning components preferred by the students, we have simulated the data that represents the characteristics of the students based on the learning styles. Table 5 shows the decision table for IFS classifier. There are 20 identified attributes that will be mapped into 16 learning styles as listed in Table 3. Attributes A1 – A6 are used to identify Active/Reflective learners, attributes A7 – A13 are the characteristics of Sensor/Intuitive learners, attributes A14 – A17 are employed to identify Visual/Verbal learners and, attributes A18 – A20 are applied to identify Sequential/Global learners.

Table 5. Decision table for IFS classifier

A1	A2	A3	A4	A5	A6	A7	A8	A9	A10	A11	A12	A13	A14	A15	A16	A17	A18	A19	A20	Decision
1	1	1	1	1	1	1	1	0	1	1	0	0	0	1	0	1	0	1	1	ASViSq
1	1	1	1	1	0	1	1	0	1	1	0	1	0	0	0	1	0	0	0	ASViSq
1	1	1	1	0	0	1	1	0	1	1	1	0	1	1	1	1	0	1	0	ASViSq
1	1	1	1	0	1	1	1	0	1	1	1	1	1	1	1	0	0	1	1	ASViSq
0	1	1	1	0	1	1	1	0	1	1	0	0	0	1	0	1	0	1	1	RSViSq
0	1	1	0	1	1	1	1	0	1	1	0	1	0	0	0	1	0	0	0	RSViSq
0	1	1	0	0	1	1	1	0	1	1	1	0	1	1	1	1	0	1	0	RSViSq
1	1	1	1	1	1	1	0	1	1	1	0	0	0	1	0	1	0	1	1	AIViSq
1	1	1	1	1	0	1	0	1	1	1	0	1	0	0	0	1	0	0	0	AIViSq
1	1	1	1	0	0	1	0	1	1	1	1	0	1	1	1	1	0	1	0	AIViSq
1	1	1	1	0	1	1	0	1	1	1	1	1	1	1	1	0	0	1	1	AIViSq
0	0	0	1	0	1	1	0	1	1	0	1	0	0	0	1	1	0	0	1	RIVbG
0	0	0	0	1	1	1	0	1	1	0	1	1	0	0	1	0	1	1	1	RIVbG
0	0	0	0	1	0	1	0	1	0	1	0	0	0	1	1	1	1	0	1	RIVbG
0	0	0	0	0	1	1	0	1	0	1	0	1	0	1	1	0	1	1	0	RIVbG

7 Experimental Results

The analysis of the proposed study is validated by executing rough set rules measurement. Rules are generated from reducts. The rules may be of different types and on different formats, depending on the used algorithms. A decision rule can be denoted as $\alpha \rightarrow \beta$ which implies if α then β.

Rule measurement has three types [40]:

1. **Support measurement:**
Given a description contains a conditional part, α and the decision part β, denoting a decision rule $\alpha \rightarrow \beta$. The support of the pattern α is a number of objects in the information system A has the property described by α.

$$Support(\alpha) = \|\alpha\| \tag{1}$$

The support of β is the number of the object in the information system A that have the decision described by β.

$$Support(\beta) = \|\beta\| \tag{2}$$

The support for the decision rule $\alpha \rightarrow \beta$ is the probability of that an object covered by the description is belongs to the class.

$$Support(\alpha \rightarrow \beta) = \|\alpha \cdot \beta\| \tag{3}$$

2. **Accuracy measurement:**
Presented as the quantity accuracy of ($\alpha \rightarrow \beta$) that gives a measure of how reliable the rule is in the condition of (β). It is the probability that an arbitrary object covered by the description belongs to the class. It is identical to the value of rough membership function applied to an object x that match α. Thus accuracy measures the degree of membership of x in X using attribute B.

$$Accuracy(\alpha \rightarrow \beta) = \frac{Support(\alpha \cdot \beta)}{Support(\alpha)} \tag{4}$$

3. **Coverage measurement:**
Defined as a measurement of how well the pattern α describes the decision class defined through β. It is the probability that an arbitrary object, belonging to the class C is covered by the description D.

$$Coverage(\alpha \rightarrow \beta) = \frac{Support(\alpha \cdot \beta)}{Support(\beta)} \tag{5}$$

The rules are completed if any object belonging to the class is assigned with the coverage of 1. While deterministic rule are rules with the accuracy of 1, and the correct rules are rules with both coverage and accuracy equal to 1.

7.1 Rules Analysis of the Integrated Felder Silverman Learning Style

We designed 1184 simulated data based on the attributes for IFS dimensions shown in Table 5. 80% of the data is used for training and 20% for testing. In our study, ROSETTA tool [44] has been used for classification. To date, ROSETTA has been used successfully in data analysis in various applications since it provides various discretization, reduction and classification techniques.

In the analysis phase, extensive experiments have been conducted using various discretization techniques such as Boolean reasoning, Equal Frequency Binning, Entropy/MDL algorithm and Naive algorithm. (Please refer to [38] and [45] for further explanations on discretization technique. It is essential to choose an appropriate discretization method since performance of discretization methods differ significantly [46]. The experiments were conducted in order to choose the most significant discretization method for our data.

The discretized data are further processed to mine the significant rules using Genetic Algorithm and Johnson Algorithm. ROSETTA provides two options on discernibility; full discernibility and object related discernibility. With full discernibility,

Table 6. The testing result for various choices of discretization and reduction technique

Method of discretization	Method of reduction	No. of reducts	No. of rules	Testing accuracy
Boolean reasoning	Genetic algorithm (full reduct)	61	34710	84.81
	Johnson algorithm (full reduct)	1	546	83.96
	Genetic algorithm (object reduct)	76238	84854	90.72
	Johnson algorithm (object reduct)	426	505	91.56
Equal Frequency Binning	Genetic algorithm (full reduct)	61	34710	84.81
	Johnson algorithm (full reduct)	1	546	83.96
	Genetic algorithm (object reduct)	**76292**	**85253**	**92.41**
	Johnson algorithm (object reduct)	426	505	91.56
Entropy/MDL algorithm	Genetic algorithm (full reduct)	61	34710	84.81
	Johnson algorithm (full reduct)	1	546	83.9662
Naive algorithm	Genetic algorithm (full reduct)	46	26918	83.122
	Johnson algorithm (full reduct)	1	546	83.9662

a set of minimal attribute subsets that can distinguish all objects from each other is produced. With object related discernibility, reducts that can differentiate a certain object from others are generated. We used both options of discernibility and classified the data using standard voting classifier. A total of 12 different approaches have been conducted in classifying the data using rough set classifier.

Table 6 reveals the results of the experiments, and it shows that all approaches have more than 80% testing accuracy. The best testing accuracy is 92.41% which is achieved through Equal Frequency Binning and Genetic Algorithm for object reduction. Equal frequency binning usually generates high testing accuracy compare to other methods since it creates discretization intervals; an equal number of objects in each interval. Genetic algorithm reduction performed better than Johnson's algorithm since it provides a more exhaustive search [41].

Subsequently, the classifier performance with Equal Frequency Binning discretization and Genetic algorithm reduction is examined. The simulated data is randomly divided into 4 training and testing set with a partition of 80%, 70%, 60% and 50% accordingly. The classification results are depicted in Table 7. Higher consumption of data at training phase gives larger number of rules; hence yields better accuracy. The classifier with 80% training data and 20% testing data furnishes the highest accuracy and the largest number of generated rules.

Table 7. Comparison of classification with various partition of training and testing data

% Training Data	% Testing Data	RHS Support	LHS Length	Number of Rules	Testing Accuracy
80	**20**	**1-43**	**4-15**	**85253**	**92.41**
70	30	1-39	4-16	80827	82.54
60	40	1-36	4-14	74832	78.90
50	50	1-32	4-14	66700	73.3

Consequently, for better accuracy, the partition of 80% training data and 20% testing data is chosen, and the rules generated by this classifier are further examined for filtering and mining the most significant rules. Example of the generated rules is shown in Table 8. Each generated rule is associated with the right hand side (RHS) support and left hand side (LHS) support, accuracy, coverage, and length. The definition of the rule statistics are as below [41]:

1. LHS SUPPORT – the number of records in the training data that fully exhibit the property described by the IF-THEN conditions.
2. RHS SUPPORT – the number of records in the training data that fully exhibit the property described by the THEN conditions.
3. RHS ACCURACY – the number of RHS support divided by the number of LHS support.
4. LHS COVERAGE – the fraction of the records that satisfied the IF condition of the rule. It is obtained by dividing the support of the rule by the total number of records in the training sample.

5. RHS COVERAGE - the fraction of the records that satisfied the THEN conditions. It is obtained by dividing the support of the rule by the total number of records in the training sample that satisfied the THEN CONDITION.
6. LENGTH – the number of conditional elements in the IF part.

From the experiment conducted, rough set method generates a large number of reduct and rules, with 76292 reducts and 85253 rules. Reducts are used to generate decision rules. A reduct is able to generate several rules and it can be seen from the experiment that the classifier produce 85253 rules from 76292reducts generated. Table 8 extracts several rules with higher rule support and lowest rule support. The decision rule, at the left side, is a combination of values of attributes such that the set of all objects matching this combination have the decision value at the rule's rough side. The rule derived from reducts can be used to classify the data. The set of rules is referred as classifier and can be used to classify new and unseen data.

Table 8. Sample of the rules generated by rough set

Rules	LHS support	RHS support	RHS accuracy	LHS coverage	RHS coverage	LHS length	RHS length
PRForum(1) AND Abstract(0) AND Simulation1(0) AND HtextCov1(1) AND Linear/global(1) => Label(ASVbG	43	43	1	0.045407	0.614286	5	1
PRForum(1) AND Abstract(1) AND Simulation1(1) AND Linear/global(1) => Label (AIViG)	35	35	1	0.036959	0.5	4	1
PRForum(1) AND Abstract(0) AND Simulation1(1) AND Linear/global(0) AND Courseview(0) => Label(ASViSq)	32	32	1	0.033791	0.484848	5	1
PRForum(1) AND Abstract(0) AND Simulation1(1) AND Linear/global(0) AND HTextCov2(1) => Label(ASViSq)	29	29	1	0.030623	0.439394	5	1
PRForum(0) AND HtextCov(1) AND Concrete(1) AND HtextCov1(0) AND Linear/global(0) AND Courseview(0) => Label (RSViSq)	14	14	1	0.014784	0.28	6	1
PRForum(0) AND Simulation(0) AND codeExec(0) AND HtextCov(0) AND ExmRvsion(0) AND HtextCov1(1) AND pptAccess(1) AND Linear/global(1) AND HTextCov2(0) => Label(RIVbG)	1	1	1	0.00105597	0.0204082	9	1
PRForum(0) AND Simulation(1) AND codeExec(1) AND forumView(0) AND HtextCov(1) AND Abstract(1) AND Exmple(1) AND Exmdlvry(0) AND ExmRvsion(1) AND HtextCov1(1) AND HTextCov2(0) AND Courseview(1) => Label(RIVbG)	1	1	1	0.00105597	0.0204082	12	1

When rules are generated, the numbers of objects that generate the same rule is typically recorded and represent LHS rule support. The highest LHS rule support from this experiment is 43 and the lowest LHS support is 1. Meanwhile the minimum LHS length of the conditional elements in the generated rules is 4 and the maximum length of the rules is 15. The accuracy of the rules is 1 since there is no inconsistency in the decision system (inconsistency rule is the rule that has several decisions in the THEN part). The value of rule support, rule coverage and accuracy are computed based on equation (3), equation(4) and equation (5).

However, not all rules are important and it is vital to choose the rules that are significant in the classification process. Due to the large number of rules, it is difficult to analyze these rules manually. The following section will discuss the rule filtering approach in order to obtain the granularity of the significant features.

7.2 *Rule Filtering*

Rule filtering involves the process of eliminating insignificant rules from the generated rule sets. The criteria of the filtering process are as follows:

1. Filtering based on the left hand side (LHS) length.
2. Filtering based on the left hand side (LHS) support.
3. Overall testing accuracy

The rule filtering entails the stepwise elimination of insignificant rules based on the criteria mentioned above.

7.2.1 Filtering the Rule Length

Table 9 illustrates the experimental results on the impact of filtering the rules based on various rule length. Fig. 7 exhibits the classification accuracy of reducing the rules based on the selected length. Rule filtering is done by continuously

Table 9. Classification accuracy with various LHS rule length

LHS length	No. of rules	Testing accuracy
4-15	85253	92.41
4-14	85237	92.41
4-13	85078	92.41
4-12	84219	92.41
4-11	81015	92.41
4-10	71744	92.41
4-9	55180	92.41
4-8	**35764**	**92.41**
4-7	18398	91.56
4-6	5673	85.23
4-5	498	51.05

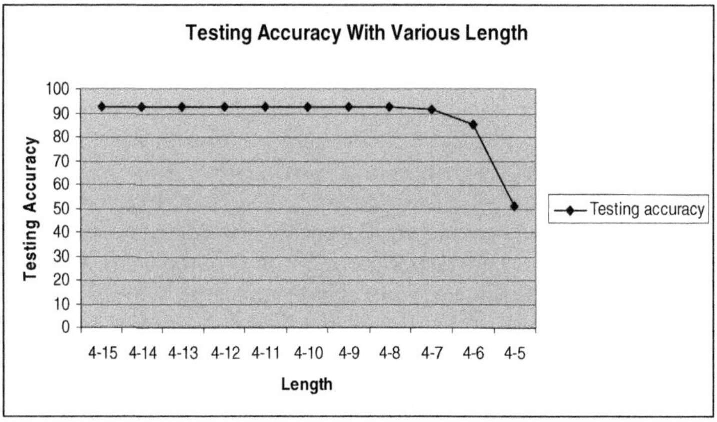

Fig. 7. Classification accuracy with various rule LHS length

eliminating the rule with the highest length and so forth. The result shows that the classification accuracy is similar despite rules reduction from the maximum LHS length of 15 to LHS length of 8. However, by filtering the rule from LHS of length 7 to 5, the accuracy rate decreases tremendously. This implies that the rules with shorter length are more important than the longest length. Subsequently, further filtering process is executed by opting for the rules with LHS of length 4 to LHS length 8.

7.2.2 Filtering Based on the Rule Support

Table 10 demonstrates the results of filtering the rules based on various LHS support. In this experiment, we filter the rules by eliminating the rules with the minimum support starting with LHS support 1 until LHS support 15. The results reveal that the classification accuracy gradually increases by eliminating the rule with minimum LHS support of 2 to minimum LHS support of 5. The rules within the range of 6-43 give the highest accuracy, 96.20%. Meanwhile, the accuracy rate decreases by eliminating the LHS support from 10 onwards. The result implies that the rules with lower LHS support are non-essential and doesn't have effect on the accuracy, unlike rules with higher support of 6 – 43, which are really vital for higher accuracy.

We further choose the rules with the length between 4-8 and the support in the range of 1-43 to 10-43 for additional classification. Table 11 shows that rules with LHS support of 5- 43 give the highest accuracy, followed by the rule with LHS support of 6-43 and 7-43. To decide on the best range of the rule support for the classifier, we compute the average accuracy for three classifiers which is 96.6%. The classifier with the rule support of 6-43 has the accuracy of 96.62 and 4290 rules in this range. The rules generated in this classifier are only 5% from the initial set of rules generated by rough set.

Table 10. Classification accuracy with various LHS support

Genetic algorithm with object reduct	**LHS support**	**No. of rules**	**Testing accuracy**
All rules	1-43	85253	92.41
	2-43	65109	92.41
	3-43	24008	94.1
	4-43	16106	94.94
Rules with minimum support < 10	5-43	7441	95.78
	6-43	5063	96.20
	7-43	3086	95.78
	8-43	2163	92.83
	9-4	1439	87.76
	10-43	1041	83.96
	11-43	732	78.48
Rules with minimum support >= 10	12-43	546	74.68
	13-43	432	70.88
	14-43	335	62.45
	15-43	284	59.49

Table 11. Classification accuracy for rules with various support and LHS length between 4-8

Rule support	Testing accuracy	No of rules
1-43	92.41	35764
2-43	93.25	28991
3-43	94.94	14868
4-43	94.1	10843
5-43	**97.47**	**6051**
6-43	**96.62**	**4290**
7-43	**95.78**	**2792**
8-43	91.98	2004
9-43	87.34	1370
10-43	84.81	1003

Fig. 8 shows the performance comparison of two classifiers using the same rule support but different rule length. The rule length being considered in this study is (4-15) which is the rule length of the original rules generated, and (4-8) is the range of the rule length that gives the highest accuracy (Table 9). The comparison result illustrates that even though the rule length has been reduced; as a whole, the accuracy rate is not much affected compared to the classification rates with the initial rule length. However, for some range of rule support, such as 5- 43 and 6-43, the rules with reduced length are able to give higher accuracy.

Fig. 9 shows the comparison of number of generated rules which has the same rule support but different rule length. The rule length being considered in this study is (4-15) which is the rule length of the original generated rules and (4-8) is the range of the rule length that gives the highest accuracy as depicted in Table 9. The result shows that by reducing various rule support, the number of rules in both range of length (4-15 and 4-8) is decreasing. However, the rules with the length of (4-8) has fewer number of rules compared with the rules with the length of (4-15). The results illustrated that most of the initial rules generated by rough sets were redundant and unimportant. By filtering the rules based on the rule support and rule length, we are able to extract the most significant rule for better accuracy.

Table 12 provides the comparison of implementing Neural Network (NN) and rough set. Rough set with filtered rules are much better in term of classification accuracy and number of rules. It is observed that rough set is able to find the most significant rules by filtering the rule length and rule support. Meanwhile, the

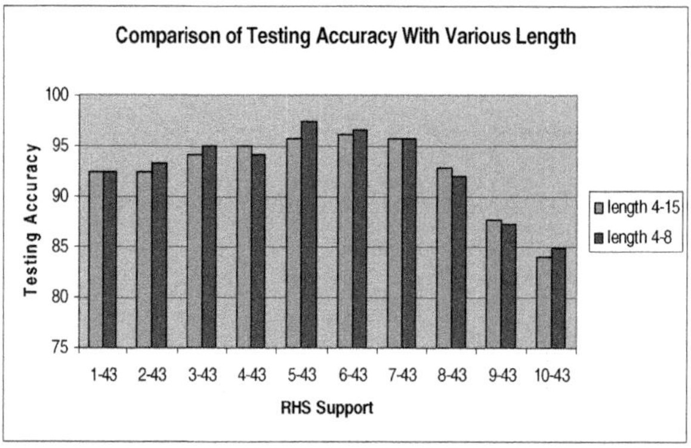

Fig. 8. Comparison of testing accuracy with various rule length

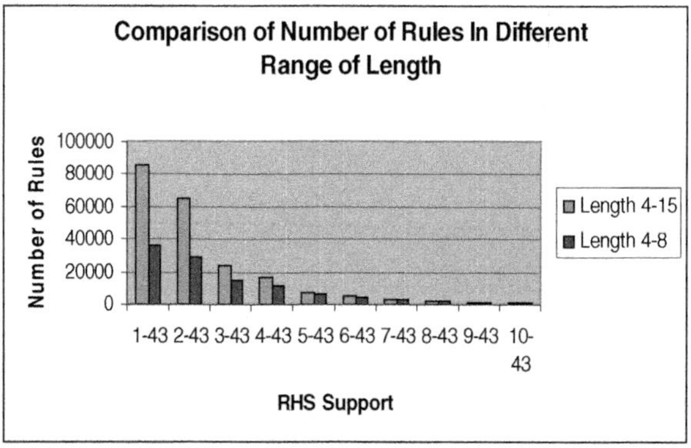

Fig. 9. Comparison of number of rules in different range of rule length

experiments with NN showed that the network is able to classify the learning dimension of a student by examining the students' interaction in the hypermedia learning system. The results showed that NN performed well in identifying the learning styles; however Rough Sets with reduced rules give the highest accuracy. The results revealed that the filtered rules give successful classification rate.

Table 12. Comparison of the classification accuracy result

Algorithm	Classification Accuracy
Rough Set with initial generated rules - 85253 rules	92.41
Rough Set with filtered rules -4290 rules	96.62
Neural Network	94.75

8 Conclusion and Future Work

In this study, we proposed an integrated Felder Silverman learning style model by analyzing student's preferences while using e-learning system developed using Moodle. The flexibility of Moodle allows developing an e-learning system that incorporates various learning resources based on Felder Silverman learning style model. We have identified significant characteristics related to Felder Silverman learning dimension: active/reflective, sensor/intuitive, visual/verbal and sequential/global. We found out that the preferences of the students are consistent with the characteristics of the learning styles describe in Felder Silverman model. These outcomes can be benefited by educators who wish to incorporate various learning material in their e-learning presentations by associating the content with the student's learning style.

From the conducted experiments, we conclude that rough set is able to correctly classify learner characteristics into 16 integrated Felder Silverman learning dimensions. Initially, rough set generate excessive rules associated with each class. However, we mine the granularity of these rules by identifying the most significant rules for better accuracy. This is achieved by filtering the rules. These rules are reduced based on the rule support and the rule length. From our experiments, we found that the rules with higher support and minimum length are much significant compared to the rules with less support and minimum length. The results depict that the classification is better with highest rule support and minimum rule length, and only 5% from the initial set of rules generated by rough set is significant.

For future, we are going to extend the research by incorporating a recommendder system in Moodle. Such system will recommend the learning materials and learning tasks based on the IFS learning style. The classifier will be integrated into Moodle in order to diagnose the learning behavior of a student and map the behavior into IFS. Data extracted from Moodle log must be preprocessed and fed to the classifier to identify the student's learning style. The information is crucial in order to develop a recommender system in LMS.

References

[1] Kerdprasop, N., Muenrat, N., Kerdprasop, K.: Decision Rule Induction in a Learning Content Management System. Proceedings of World Academy of Science, Engineering and Technology 2(2), 77–81 (2008)

[2] Moodle (2009), `http://moodle.org/` (Retrieved May 15, 2009)

[3] Blackboard (2009), `http://www.blackboard.com/` (Retrieved May 15, 2009)

[4] WebCT (2007), `http://www.webct.com/` (Retrieved April 30, 2007)

[5] Graf, S.: Adaptivity in Learning Management System Focusing on Learning Styles. Phd Thesis, Vienna University of Technology (2007)

[6] Romero, C., Ventura, S., Garcia, E.: Data Mining in Course Management Systems: Moodle Case Study and Tutorial. Computers & Education (51), 368–384 (2008)

[7] Mostow, J., Beck, J.: Some Useful Tactics To Modify, Map and Mine from Data from Intelligent Tutors. Natural Language Engineering 12(2), 195–208 (2006)

[8] Ai, J., Laffey, J.: Web Mining as a Tool for Understanding Online Learning. MERLOT Journal of On-line Learning and Teaching 3(2), 160–169 (2007)

[9] Wolf, C.: iWeaver: Towards an Interactive Web-Based Adaptive Learning Environment to Address Individual Learning Styles. In: Proceedings Fifth Australasian Computing Education Conference (ACE 2003), Adelaide, Australia, pp. 273–279 (2003)

[10] Papanikolau, K., Grigoriadou, M., Knornilakis, H., Magoulas, G.: Personalizing the Interaction in a Web-based Educational Hypermedia System: the case of INSPIRE. User Modeling and User-Adapted Interaction (13), 213–267 (2003)

[11] Carver, C.A., Howard, R.A., Lavelle, E.: Enhancing Student Learning by Incorporating Learning Styles into Adaptive Hypermedia. In: Proceedings of ED-MEDIA 1996 World Conference on Educational Multimedia Hypermedia, pp. 118–123 (1996)

[12] Triantafillou, E., Pomportsis, A.A., Georgiadou, E.: AES-CS: Adaptive Educational System based on Cognitive Styles. In: Second International Conference on Adaptive Hypermedia and Adaptive Web-based Systems, Malaga, Spain, May 29-31 (2002)

[13] Villaverde, J., Godoy, D., Amandi, A.: Learning Styles' Recognition In E-Learning Environments With Feed-Forward Neural Networks. Journal of Computer Assisted Learning 22(3), 197–206 (2006)

[14] Stash, N., de Bra, P.: Incorporating Cognitive Styles in AHA (The Adaptive Hypermedia Architecture). In: Proceedings of the IASTED International Conference Web-Based Education, Austria, pp. 378–383 (2004)

[15] Kelly, D., Tangney, B.: Predicting Learning Characteristics in a Multiple Intelligence Based Tutoring System. In: Lester, J.C., Vicari, R.M., Paraguaçu, F. (eds.) ITS 2004. LNCS, vol. 3220, pp. 678–688. Springer, Heidelberg (2004)

[16] García, P., Amandi, A., Schiaffin, S., Campo, M.: Evaluating Bayesian Networks Precision for Detecting Students' Learning Styles. Computers & Education 49, 794–808 (2007)

[17] Graf, S., Kinshuk: An Approach for Detecting Learning Styles in Learning Management Systems. In: Proceedings of the International Conference on Advanced Learning Technologies, pp. 161–163. IEEE Computer Science, Los Alamitos (2006)

[18] Yaannibelli, V., Godoy, D., Amandi, A.: A Genetic Algorithm Approach to Recognize Students' Learning Styles. Interactive Learning Environments 14(1), 55–78 (2006)

[19] Lo, J., Shu, P.: Identification Of Learning Styles Online by Observing Learners' Browsing Behaviour Through A Neural Network. British Journal of Educational Technology 36(1), 43–55 (2005)

[20] Cha, H., Kim, Y., Lee, J., Yoon, T.: An Adaptive Learning System With Learning Style Diagnosis Based On Interface Behaviors. In: Workshop Proceedings of International Conference on E-Learning and Games, Hangzhou, China, pp. 513–524 (2006)

[21] Ahmad, N., Shamsuddin, S.M.: Mapping Student Learning Characteristics Into Integrated Felder Silverman Learning Style Using Neural Network. In: Proceedings of 1st International Malaysian Educational Technology Convention, Malaysia, Malaysia, November 2-5, pp. 202–209 (2007)

[22] Mak, B., Manakata, T.: Rule Extraction from Expert Heuristics: A Comparative Study of Rough Sets With Neural Networks and ID3. European Journal of Operational Research 136, 212–229 (2002)

[23] West, W., Rosser, B.R.S., Monani, S., Gurak, L.: How Learning Styles Impact E-Learning: a Case Comparative Study Of Undergraduate Students Who Excelled, Passed Or Failed An Online Course In Scientific/Technical Writing. E-learning 3(4), 534–543 (2006)

[24] Garcia, E., Romero, C., Ventura, S., Calders, T.: Drawbacks and Solutions of Applying Asoociation Rule Mining in Larning Management Systems. In: Proceedings of the International Workshop on Applying Data Mining in e-Learning (ADML 2007), Crete, Greece, pp. 13–22 (2007)

[25] Graf, S., Kinshuk: Analysing the Behavior of Students in Learning Management Systems with Respect to Learning Styles. Studies in Computational Intelligence (SCI) 93, 53–73 (2008)

[26] Felder, R., Brent, R.: Understanding Student Differences. Journal of Engineering Education (1), 57–72 (2005)

[27] Kolb, D.A.: Experiential Learning: Experience as the Source of Learning and Development. Prentice-Hall, Englewood Cliffs (1984)

[28] Gardner, Howard: Multiple Intelligences: The Theory in Practice. Basic, New York (1993)

[29] Honey, P., Mumford, A.: A Manual of Learning Styles. Peter Honey, Maidenhead (1986)

[30] Dunn, Dunn: Teaching Students Through Their Individual Learning Styles: A Practical Approach. Reston Publishing Company, Inc., Virginia (1978)

[31] Zywno, M.S.: A Contribution to Validation of Score Meaning for Felder-Soloman's Index of Learning Styles. In: Proc. of the 2003 American Society for Engineering Annual Conference and Exposition (2003), `http://www.ncsu.edu/felderpublic/ILSdir/Zywno_Validation_Study.pdf` (Retrieved August 2, 2006)

[32] Carmo, L., Marcelino, M., Mendes, A.: The Impact of Learning Styles in Introductory Programming Learning, International Conference on Engineering Education-ICEE 2007, Coimbra, Portugal (2007), `http://icee2007.dei.uc.pt/proceedings/papers/432.pdf` (Retrieved September 9, 2009)

[33] Felder, R., Silverman, L.: Learning And Teaching Styles In Engineering Education. Engineering Education 78(7), 674–681 (1988), `http://www.ncsu.edu/felderpublic/Papers/LS-1988.pdf` (Retrieved September 9, 2009)

[34] Felder, R.M., Solomon, B.A.: Index of Learning Styles Questionnaire (2009), http://www.engr.ncsu.edu/learningstyles/ilsweb.html (Retrieved May 15, 2009)

[35] Bajraktarevic, N., Hall, W., Fullick, P.: Incorporating Learning Styles in Hypermedia Environment: Empirical evaluation. In: Proceedings of AH2003: Workshop on Adaptive Hypermedia and Adaptive Web-based Systems, Budapest, Hungary, pp. 41–52 (2003)

[36] Ahmad, N., Shamsuddin, S.M.: Analyzing Learning Preferences From E-learning Activities. In: E-learning Technology and Applications. UTM Press (2008)

[37] Pawlak, Z.: Rough Sets: Theoretical Aspect of Reasoning About Data. Kluwer Publications, Dordrecht (1991)

[38] Swiniarski, R.: 'Rough Sets Methods in Feature Reduction and Classification'. International Journal Appli. Math Computer Science 11(3), 565–582 (2001)

[39] Li, J., Cercone, N.: Introducing a Rule Importance Measure. In: Peters, J.F., Skowron, A. (eds.) Transactions on Rough Sets V. LNCS, vol. 4100, pp. 167–189. Springer, Heidelberg (2006)

[40] Øhrn, A.: Discernability and Rough Sets in Medicine: tools and applications, PHD Thesis. Department of Computer and Information Science, Norwegian University of Science and Technology, Norway (1999), http://www.idi.ntnu.no/~aleks/thesis

[41] Bose, I.: Deciding the Financial Health of Dot-coms using Rough Sets. In: Information and Management, pp. 835–846 (2006)

[42] Abidi, S.S.R., Hoe, K.M., Goh, A.: Analyzing Data Clusters: A Rough Set Approach To Extract Cluster-Defining Symbolic Rules. In: Hoffmann, F., Adams, N., Fisher, D., Guimarães, G., Hand, D.J. (eds.) IDA 2001. LNCS, vol. 2189, pp. 248–257. Springer, Heidelberg (2001)

[43] Hor, C.L., Crossley, P.A., Watson, S.J.: Building Knowledge for Substation-Based Decision Support Using Rough Sets. IEEE Transactions on Power Delivery 22(3), 1372–1379 (2007)

[44] Øhrn, A.: ROSETTA homepage, http://rosetta.lcb.uu.se/general/ (retrieved December 2007)

[45] Liu, H., Husain, F., Tan, C.L., Dash, M.: Discretization: An Enabling Technique. In: Data Mining and Knowledge Discovery, vol. 6, pp. 393–423. Kluwer Academic Publishers, Dordrecht (2002)

[46] Blajdo, P., Grzymala-Busse, J.W., Hippe, Z.S., Knap, M., Mroczek, T., Piatek, L.: A Comparison of Six Discretization – A Rough Set Perspective. In: Wang, G., Li, T., Grzymala-Busse, J.W., Miao, D., Skowron, A., Yao, Y. (eds.) RSKT 2008. LNCS (LNAI), vol. 5009, pp. 31–38. Springer, Heidelberg (2008)

T-Learning 2.0: A Personalised Hybrid Approach Based on Ontologies and Folksonomies

Marta Rey-López, Rebeca P. Díaz-Redondo
Ana Fernández-Vilas, and José J. Pazos-Arias

Dept. of Telematic Engineering, University of Vigo, 36310, Spain
{mrey,rebeca,avilas,jose}@det.uvigo.es

Abstract. Collaborative tagging is becoming a popular practice to annotate resources on the web, which has even reached e-learning initiatives. The aim of this chapter is exporting this technology to the field of learning through Interactive Digital TV (IDTV) (t-learning). In previous research, we have exposed our solution for creating learning experiences for t-learning, based on combining TV programmes and learning elements in order to lure viewers into education and make these experiences more entertaining. At the beginning of this research, we suggested reasoning over ontologies for the combination of the different elements. However, this approach did not take into account the user's point of view towards the contents. In this chapter, we go a step further and present a proposal that includes collaborative tagging techniques, complementing ontologies with folksonomies to establish the relationships between the contents linked to create the learning experiences.

1 Introduction

In the last few years, governments have witnessed the success of distance courses for long-life learning. Hence, they are empowering initiatives to make easier the access to these courses for all social classes. In this scenery, Interactive Digital TV (IDTV) plays a paramount role, since the vast majority of the population of developed and developing countries has access to this medium. Several initiatives have arisen to make t-learning (learning through IDTV) become a reality [8]. However, most of these initiatives do not take into account the particular characteristics of t-learners, who are viewers and learners at the same time.

In order to solve this issue, we have presented a solution that offers personalised learning through IDTV using TV programmes to lure viewers into education, called *entercation* (entertainment that educates) [9]. It consists in offering the subscribers learning objects that may be of their interest, related to the issues dealt with in the TV programme they are watching.

For example, imagine an episode of the series 'Grey's Anatomy' where the alcoholism of the main character's father gets worse, a patient who has a swastika

F. Xhafa et al. (Eds.): Computational Intelligence for Tech. Enhanced Learning, SCI 273, pp. 125–142.
springerlink.com

tattooed refuses to be treated by a doctor of a different race than his and the doctors have to perform a craniotomy to remove a brain tumor. The viewers could be offered related learning elements —such as a documentary about preventing alcoholism, a lesson about the holocaust or an educational object about different types of brain tumors— that would help them understand the topics or get deeper into them. However, for particular users, the system should only select those elements which are interesting to them and appropriate for their knowledge in order not to overwhelm them. For example, the lesson about the holocaust should only be offered to people who likes learning History.

To automatically create these *entercation* experiences, it is necessary to establish relationships between traditional TV programmes and learning objects and filter these relationships according to the learners' interests and learning background. In this chapter, we provide a solution for finding these relationships that combines two well-known tools in the semantic web that are usually considered as flip-sides of the same coin: ontologies —formal structures created a priori for experts in the subject— and folksonomies —fuzzy structures created from the collaboration of multiple users in providing descriptions for the shared contents (web pages, pictures, blog entries, movies, etc.).

2 Related Work

In previous research [9], we have exposed a proposal where we established the aforementioned relationships by means of ontologies. An ontology is a rigorous and exhaustive organization of some knowledge domain which is usually hierarchical and contains all the relevant relationships between the concepts of this domain. In our proposal, we used three interrelated ones: two of them based on the metadata that describe the contents (TV-Anytime [12] for TV programmes and ADL SCORM [1] for learning elements) and a third one —called gateway ontology— that describes the subject and allows to relate the other two. However, this method, apart from being too heavy in terms of computing power, is based on the fact that content creators describe the contents accurately and appropriately. But this hypothesis is not usually true. Besides, it only takes into account the creators' point of view, which is usually very different from the users' one.

With the aim of improving this solution, we have taken into consideration the new phenomena arisen in the so-called Web 2.0 where different users collaborate in creating, sharing and describing contents. One of these phenomena is collaborative tagging, used in web sites such as Del.icio.us[1] or Flickr[2] , which allows users to tag and share contents so as they cannot only categorize —assigning tags to it— their own content but also the one incorporated by others [3].

In this manner, each piece of content is assigned a set of tags with their respective weights, which are proportional to the number of users who have described the particular content with this tag. This is called a tag cloud and is graphically depicted

[1] http://del.icio.us
[2] http://www.flickr.com

as shown in Fig. 1(a), with more popular tags represented in a bigger font size. Similarly, the users of the system have their own tag cloud, containing the tags they have ever assigned with a weight that is proportional to the number of times they have used this tag to describe any content (see Fig. 1(b)).

(a) Grey's Anatomy tag cloud (b) User's tag cloud

Fig. 1. Tag clouds

Using the contents' tag clouds, obtained from collaborative tagging, we can generate a type of structure that Thomas Vander Wal coined in 2004 with the term folksonomy [13], defined as 'the user-created bottom-up categorical structure development with an emergent thesaurus'.

On implementation terms, a folksonomy is a structure that can be represented as an undirected graph where nodes are the different tags assigned on the system and arcs represent the relationships between the tags they link (see bottom of Fig. 5). These are weighted relationships, in a way that the more times the tags are assigned together to the same piece of content, the highest the weight of their relationship [6].

This kind of structure suggests using it to establish relationships between the different contents of the system. In fact, this idea is part of the approach presented in the current chapter, but similar approaches have been presented in the literature for finding similarities between tags [10], recommendation of movies [11] or web pages [7], as well as the calculus of the relevance of the results of a search [4].

Although ontologies and folksonomies seem to be contradictory options for the description of contents, our solution merges both of them (see Fig. 5). In this manner, we combine a folksonomy created from a collaborative tagging system for IDTV that allows the viewers to label the contents of their interest with two ontologies that describe the contents. For TV programmes, we use the TV-Anytime ontology mentioned above, while for learning elements we use an ontology that describes the domain of the subject they teach.

3 Tagging Contents

In our proposal, viewers are able to tag the content they watch, whether television programmes or educational elements, using the collaborative tagging interface

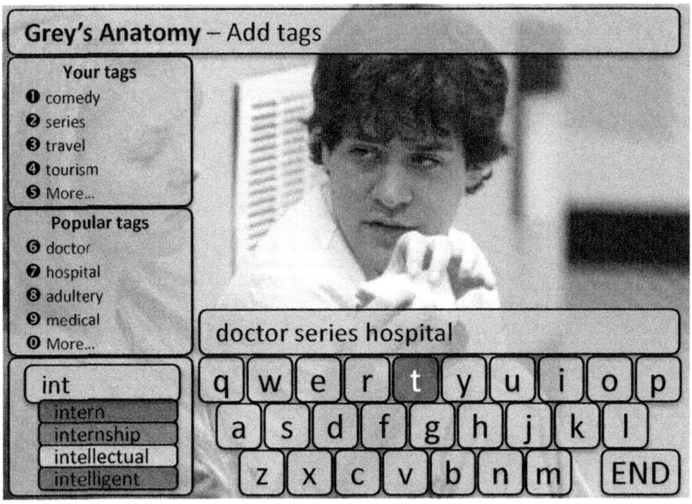

Fig. 2. Tagging interface

shown in Fig. 2. This interface was designed to be controlled by using the remote control only, since very few homes have a keyboard connected to the set-top box, at least for the time being. Given the difficulties inherent in performing this process using only this device, viewers are provided with a set of aids to make tagging easier, such as predictive text and suggestion of popular tags.

3.1 Describing the User

The tags the users choose to describe the pieces of content they watch constitute their tag cloud. The users' tag clouds are stored in their set-top box and serve as their profile. In our proposal, the weight of the tags is not only proportional to the number of times the user has assigned this tag, but also to the degree of interest (DOI) and/or knowledge (DOK) shown for the content tagged. In fact, the tags in the user's tag cloud have two different weights, one for the degree of interest and the other for the degree of knowledge of the user in this tag. Both are calculated from the degree of interest and knowledge of the user on the pieces of content the tag has been assigned to. The degree of interest towards a piece of content is deduced by the system from the proportion of time the users have watched the programme or the proportion of lessons of a learning element they have followed.

Mathematically speaking, the user profile for the user u_l consists on a set of the contents that the user has viewed or studied (denoted as C_{u_l}), as well as his/her tag cloud. This tag cloud is represented as a multiset D_{u_l} consisting on the set of tags the user has ever assigned (T_{u_l}) and two functions that relate each tag with its DOI ($DOI : T_i \rightarrow \mathbb{R}$) and DOK ($DOK : T_i \rightarrow \mathbb{R}$) shown by the user towards the content.

The interest of a user with respect to a particular tag ($DOI(t_i, D_{u_l})$) is calculated:

$$DOI(t_i, D_{u_l}) = \frac{\sum_{\forall k / c_k \in C_{u_l}(t_i)} IOP(c_k) w(t_i, c_k) DOI(c_k, u_l)}{\sum_{\forall k / c_k \in C_{u_l}(t_i)} IOP(c_k)} \quad (1)$$

Being $C_{u_l}(t_i)$ the subset of contents of C_{u_l} that the user has tagged with t_i.

The DOI of a user on a particular piece of content is a numeric value in the interval $[-1, 1]$ that reflects the interest of the user on this content, which can be explicitly given by the user or calculated from the proportion of the programme he/she has watched or the learning element he/she has estudied (Index of Viewing, IOV) in the interval $[0, 1]$ and the Index Of Feedback (IOF), which indicates the opinion of the user on the piece of content, also in the interval $[-1, 1]$. To calculate the IOV of the TV programmes, we measure the proportion of time the user has watched. For learning elements, we use the recommendations established in the ADL SCORM standard [2]. In this manner, if the element is an asset[3] we cannot exactly determine the IOV, thus, it would take the value 0 if the user accepts the content and 1 if he/she rejects it. In the case of SCOs[4], they can communicate this value to the Learning Management System (LMS)[5] using an element provided in the SCORM Data Model: *Progress Measure* —with values in the interval $[0, 1]$.

Concerning IOF, if the user does not explicitly provide it, it will be automatically assigned 0.75 if he/she accepts the content and -0.75 if he/she rejects it. In this manner, the DOI is calculated:

$$DOI(c_i, u_l) = \begin{cases} IOV(c_i, u_l) IOF(c_i, u_l) & \text{if the user accepts } c_i \\ IOF(c_i, u_l) & \text{otherwise} \end{cases} \quad (2)$$

Concerning the degree of knowledge of the user on a tag ($DOK(t_i)$), we use the same mechanism shown for the DOI:

$$DOK(t_i, D_{u_l}) = \frac{\sum_{\forall k / c_k \in C_{u_l}(t_i)} IOP(c_k) w(t_i, c_k) DOK(c_k, u_l)}{\sum_{\forall k / c_k \in C_{u_l}(t_i)} IOP(c_k)} \quad (3)$$

To calculate the degree of knowledge of a user on a learning object $DOK(c_k, u_l)$ (belonging to the interval $[-1, 1]$) we also use the recommendations established in

[3] An asset is a simple resource which does not make use of the Run-Time API defined by SCORM, therefore, it does not communicate with the Run-Time environment delivering it.

[4] A Sharable Content Object (SCO) is a resource that communicates with the delivering Run-Time Environment via the SCORM Run-Time API.

[5] A Learning Management System (LMS) refers to a system that carries out a suite of functionalities designed to deliver, track, report on and manage learning content, learner progress and learner interactions.

the ADL SCORM standard. If the element is an asset we cannot exactly determine the DOK, thus, it would take the value 0.75 if the user accepts the content and -0.75 if he/she rejects it. In the case of SCOs, they can communicate this value to the Learning Management System (LMS) using the elements provided in the SCORM Data Model: *Scaled Passing Score* (SSP) —with values in the interval $[-1,1]$—, which indicates the result that the user needs to obtain in order to get through the activity; as well as the *Scaled Score* (SS) —also with values in the interval $[-1,1]$— which indicates the result obtained by the student. The DOK is then calculated:

$$DOK(c_k,u_l) = \frac{SS(c_k,u_l) - SPS(c_k,u_l)}{2} \tag{4}$$

3.2 Describing the Contents

The pieces of content of the system, whether learning elements or TV programmes, have also their own tag cloud. Tag clouds for contents are simpler than users' ones and coincide perfectly with the description we have provided in the previous section, where the weights of the tags are proportional to the number of users that have used a particular tag to describe the content; in the following manner:

Let c_i be a piece of TV content, we denote $D_i = (T_i, m_i)$ the multiset that stores the tags that users have ever used to describe this content. Being $T_i = \{t_1, t_2, \ldots, t_{n_i}\}$ the set of tags which have been assigned to this content and $m : T_i \to \mathbb{N}$ the function that relates each tag with its multiplicity in the multiset, i.e., the number of users that have used the tag t_k to describe c_i, denoted as $m(t_k, D_i)$.

We define the function $w : T_i \to \mathbb{R}$ as the one which relates each tag with its weight in the multiset, denoted $w(t_k, D_i)$, which is calculated as follows:

$$w(t_k,D_i) = \frac{m(t_k,D_i)}{|D_i|} \tag{5}$$

Being $|D_i|$ the cardinality of the multiset, i.e., the result of adding the multiplicities of all of its tags:

$$|D_i| = \sum_{\forall k / t_k \in T_i} m(t_k,D_i)$$

Hence,

$$\sum_{\forall k / t_k \in T_i} w(t_k,D_i) = 1 \tag{6}$$

In this approach, apart from storing the tag cloud for each piece of content, it is also important to know another important parameter, the Index Of Popularity (IOP), calculated from the number of tags assigned to this content with respect to the total number of tags assigned in the system:

$$IOP(c_i) = \frac{|D_i|}{\sum\limits_{\forall k} |D_k|} \tag{7}$$

3.3 Creating the Folksonomy

A folksonomy is then created from the contents' tag clouds, such as the one shown at the bottom of Fig. 5. As mentioned before, a folksonomy is an undirected graph where the nodes are the tags ever used in the system and the arcs, the relationship between the tags they link. The relationships are calculated from the number of times the tags appear together in a tag cloud, the weights of the tags in this tag cloud, as well as the index of popularity (IOP) of those contents described by both tags at the same time.

We define the relationship between two tags (t_i and t_j) with respect to one single content c_k (denoted as $r_{ij}(D_k)$), as the geometric means between the weights of both tags in D_k.

$$r_{ij}(D_k) = \sqrt{w(t_i, D_k)\ w(t_j, D_k)} \tag{8}$$

To obtain the total relationship between t_i and t_j (r_{ij}), we should take into account their partial relationships with respect to each content and appropriately weight them according to its IOP.

$$r_{ij} = \frac{\sum\limits_{\forall k/t_i \vee t_j \in T_k} IOP(c_k)\ r_{ij}(D_k)}{\sum\limits_{\forall k/t_i \vee t_j \in T_k} IOP(c_k)} \tag{9}$$

For example, in the aforementioned figure we can see that the relationship between *hospital* and *surgery* is 0.45, higher than all the rest of relationships of the extract of the folksonomy presented. This means that they appear together with high weights in many tag clouds of contents with high IOP. This folksonomy will serve us to establish relationships between the contents to create the learning experiences, as we explain in the following section.

4 Creating t-Learning Experiences

To create *entercation* experiences, we need to establish relationships between the TV programme the user is watching and the available learning elements, rejecting those learning elements which are not appropriate for the user (whether in interest or knowledge terms) even if they are highly related to the target program.

4.1 The System

The process of creating the experiences takes place in several steps (Fig. 3). First, the contents that arrive to the users' set-top box should be filtered, in order to discard the ones that are not appropriate for them, since they may not be interesting to them or because they have not enough knowledge to understand it or too much and it would be boring for them. This task is performed by the content recommenders,

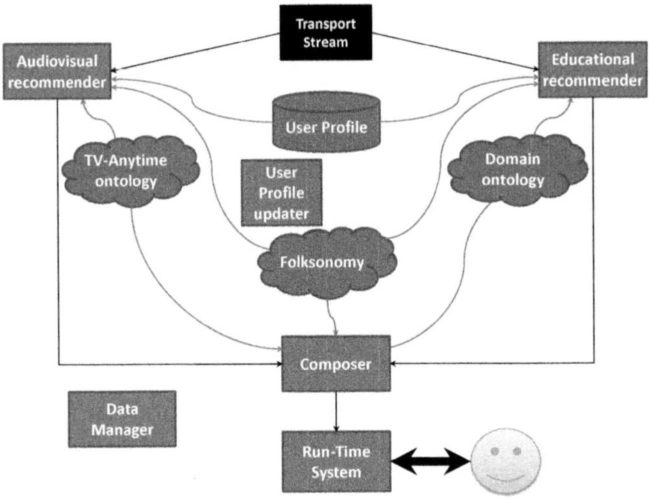

Fig. 3. Architecture of the system: creation of the experiences

both the educational and audiovisual ones, for their respective types of contents, using the algorithm to compare tag clouds that we explain in the following section.

Once the non-appropriate contents have been filtered out, the system calculates the degree of relationship between the remaining contents and selects the most compatible ones to create the learning experiences. This task is performed by the composer by using also the tag cloud comparison algorithm.

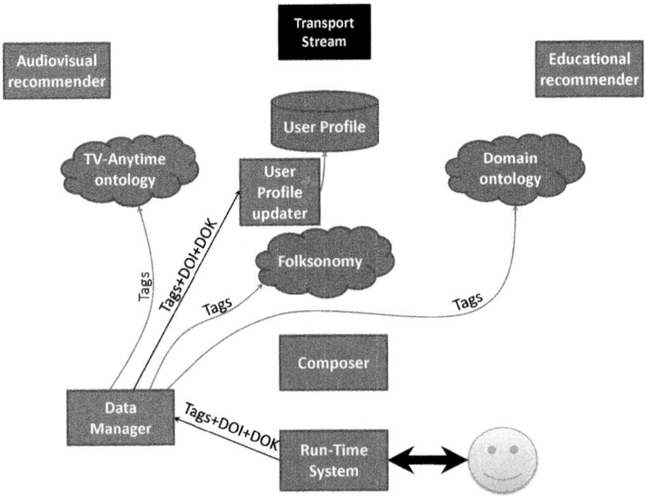

Fig. 4. Architecture of the system: feedback management

After the viewer finishes watching the programme that is the core of the learning experience, the additional learning contents are offered to him/her. The user can then decide whether to follow them or not. At any point of the programme or the learning element, the user can access the tagging interface that we have exposed in Sec. 3 with the aim of describing the content using tags, which will make easier for him/her to find the contents later and will improve the recommendations of the system.

These tags are gathered by the Data Manager (Fig. 4), as well as the degree of interest (DOI) of the user on the content and his/her degree of knowledge (DOK) on the learning element (if the content tagged belongs to this category). Hence, this system updates the folksonomy and the stereotypes of both ontologies using the new information.

4.2 The Relationships

As we have mentioned in the previous section, both the process of content recommendation and the composition of experiences need an algorithm of tag cloud comparison. In the case of content recommendation, we compare the user's tag cloud with the one of the piece of content, to establish the degree of relationship between them. On the other hand, to compose the experiences, the relationship between the different contents is obtained by comparing the tag clouds of the target contents.

The simplest way to measure this value would be to count the number of coincident tags in both tag clouds, i.e., the higher the number of coincident tags, the higher the degree of relationship between the two contents. To take into account the relative importance of the coincident tag in the tag cloud, instead of adding 1 for each coincident tag, it is better to add the means of their weights in both tag clouds. The relationships is then calculated as follows:

$$R_0(c_k, c_l) = \sum_{\forall i / t_i \in |T_k \cap T_l|} \sqrt{w(t_i, D_k)\, w(t_i, D_l)} \tag{10}$$

But this approach does not take into account the relationships between the tags. Although a tag that belongs to the first tag cloud do not appear in the second tag cloud (or appears with a lower weight), it can be closely related to the rest of the tags of the second tag cloud. For example, Fig. 5 shows an example of the calculus of the relationship between the programme 'Grey's Anatomy' (GA) and a course of 'Introduction to Cardiology' (IC). Although the tag *cardiology* has only been assigned to IC, it should contribute to increase the relationship between both contents since it is closely related to the tags assigned to GA, such as *doctor*, *hospital* or *medical*, as we can see in the folksonomy shown in the same figure.

To take this fact into account, our algorithm does not only consider the number of coincident tags and their weights (*direct relationship*, $R_0(c_k, c_l)$), but also the degree of relationship between the tags of both tag clouds (*one-hop relationship*,

$R_1(c_k, c_l)$). In this manner, the total relationship takes into account both and it is calculated

$$R(c_k, c_l) = \alpha\, R_0(c_k, c_l) + (1 - \alpha)\, R_1(c_k, c_l) \tag{11}$$

Being the parameter $\alpha \in [0, 1]$ used to assign more or less importance to the direct or one-hop relationships.

The one-hop relationship is calculated by adding the degree of relationship of each pair of possible tags (each tag of the pair belongs to one of the tag clouds) adjusted by the means of their weight.

$$R_1(c_k, c_l) = \sum_{\substack{\forall i/t_i \in T_k \\ \forall j/t_j \in T_l}} \sqrt{w(t_i, D_k)\, w(t_j, D_l)} \quad r_{ij} \tag{12}$$

The process to relate the user's tag cloud with the content's tag cloud which is used to recommend contents is slightly different. It is worth bringing back into scene the particular characteristics of the user's tag cloud, since each tag has two different weights, one related to the degree of interest and the other to the degree of knowledge. When the piece of content that we want to determine whether it is appropriate for the user or not is a TV program, we only take into account the first one. However, in the case of learning elements, we calculate the degree of interest of the user in the content in the same manner as for the TV programmes, but we also need to consider the difference between the DOK of the user in one tag and the *Degree of Difficulty* (DOD) of the learning element (which is indicated in the metadata that comes with the content).

The total relationship is then calculated as follows:

$$R(c_k, u_l) = \beta\, R_{DOI}(c_k, u_l) + (1 - \beta)\, R_{DOK}(c_k, u_l) \tag{13}$$

Being β a constant on the system which indicates whether the users give more importance to their interests or knowledge to decide the contents which should be recommended to them. In the case of TV programmes, the DOK does not have any sense, that is why β takes the value 1 for this type of contents.

These are the equations that should be used to calculate the degree of relationship between the user and the piece of content, concerning his/her interests (based on Eq. (10)(11)(12)):

$$R_{DOI}(c_k, u_l) = \alpha\, R_{DOI_0}(c_k, u_l) + (1 - \alpha)\, R_{DOI_1}(c_k, u_l) \tag{14}$$

$$R_{DOI_0}(c_k, u_l) = \sum_{\forall i/t_i \in |T_k \cap T_{u_l}|} -1^n \sqrt{w(t_i, D_k)\, |DOI(t_i, D_{u_l})|} \tag{15}$$

$$R_{DOI_1}(c_k, u_l) = \sum_{\substack{\forall i/t_i \in T_k \\ \forall j/t_j \in T_{u_l}}} -1^n \sqrt{w(t_i, D_k)\, |DOI(t_j, D_{u_l})|} \quad r_{ij} \tag{16}$$

Using n to increase or decrease the degree of relationship according to the sign of the DOI:

$$n = \begin{cases} 1 & \text{if } DOI(t_j, D_{u_l}) < 0 \\ 2 & \text{otherwise} \end{cases} \tag{17}$$

Concerning the relationship according to the DOK, we need to take into account the DOD of the content, which takes values in the interval $[-1,1]$. This value is indicated in the category `educational` $\rightarrow$ `difficulty` of the IEEE LOM [5] metadata that describe the learning element. It can take five different values: *very easy* (-1), *easy* $(-0,5)$, *medium* (0), *difficult* $(0,5)$ and *very difficult* (1). If this value is not indicated in the metadata, we assume that the DOD of content c_k is $DOD(c_k) = 0$, i.e., *medium*.

Modifying the previous equations to take into account this new factor, the degree of relationship between the user and a piece of content is calculated:

$$R_{DOK}(c_k, u_l) = \alpha\, R_{DOK_0}(c_k, u_l) + (1-\alpha)\, R_{DOK_1}(c_k, u_l) \tag{18}$$

$$R_{DOK_0}(c_k, u_l) = \sum_{\forall i / t_i \in |T_k \cap T_l|} \sqrt{w(t_i, D_k)\, \frac{1}{2}\, (2 - |DOK(t_i, D_{u_l}) - DOD(c_k)|)} \tag{19}$$

$$R_{DOK_1}(c_k, u_l) = \sum_{\substack{\forall i / t_i \in T_k \\ \forall j / t_j \in T_l}} \sqrt{w(t_i, D_k)\, \frac{1}{2}\, (2 - |DOK(t_j, D_{u_l}) - DOD(c_k)|)}\;\; r_{ij} \tag{20}$$

4.2.1 Example

Relating contents

After explaining the mechanism used to establish the relationships between contents or contents and users, let us use an example to clarify all the mathematical formulae presented. First, we will calculate the degree of relationship between an episode of the series ‘Grey’s Anatomy’ (GA) and the course ‘Introduction to Cardiology’ (IC) (Fig. 5).

Let us suppose that the tag cloud for ‘Grey’s Anatomy’ is

Tag	Weight ($w(t_k, D_i)$)
doctor	0.3
hospital	0.25
medical	0.2
intern	0.15
surgery	0.1

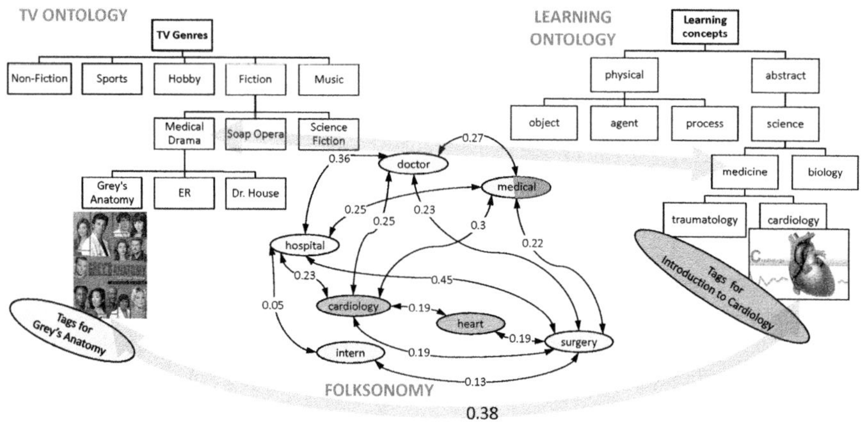

Fig. 5. Mechanism to establish the relationships between contents

For 'Introduction to Cardiology', in turn:

Tag	Weight ($w(t_k, D_i)$)
cardiology	0.5
medical	0.3
heart	0.2

Applying eq. (10) we obtain $R_0(GA, IC) = 0,24$. To calculate the one-hop relationship, we use the following table, where we can see the relationships between tags.

	$w(t_i, D_k)$ / $w(t_j, D_l)$	*doctor* 0.3	*hospital* 0.25	*medical* 0.2	*intern* 0.15	*surgery* 0.1
medical	0.5	0.27	0.25	0	0	0.22
cardiology	0.3	0.25	0.23	0.3	0	0.19
heart	0.2	0	0	0	0	0.19
$\Sigma_{\forall j/t_j \in T_i} \sqrt{w(t_i, D_k)\, w(t_j, D_l)}\ r_{ij}$		0.18	0.15	0.073	0	0.11
$R_1(GA, IC)$		0.51				

Thus, $R_1(GA, IC) = 0.51$, which makes the total relationship $R(GA, IC) = 0.38$, using $\alpha = 0.5$, which makes the direct relationship as important as the one-hop relationship.

Let us compare this value with the one obtained by calculating the degree of relationship between 'Grey's Anatomy' and a course of 'Labour Risk Prevention' (LRP) whose tag cloud is: *work* (0.4), *risk* (0.3), *disease* (0.3). We should expect this course to be less related to GA than the cardiology one, since risk prevention is related to medicine but in a lower extent than cardiology.

In this case $R_0(GA, LRP) = 0$, since they do not have any coincident tags in their tag cloud. Repeating the previous calculus, $R_1(GA, LRP) = 0.36$ so that the total relationship turns to be $R(GA, LRP) = 0.18$. Hence, it is lower than the one between the series and the course of cardiology, as expected. For this reason, when creating an educational experience, the latter is preferred to complement 'Grey's Anatomy'.

Recommending contents

To illustrate the process of content recommendation, let us expose an example for a user with the following tag cloud:

Tag	DOI	DOK
medical	0.9	0.2
football	-0.2	0.5
sport	0.7	0.6
nature	0.5	0.65

If we compare this user's tag cloud with the one of 'Grey's Anatomy', we only need to take into account the DOI, since DOK is only important for learning elements. Applying the eq. (15), the direct relationship is $R_{DOI_0} = 0.42$.

In turn, the one-hop relationship can be calculated using this table:

	$DOI(t_j, D_{u_l})$ \ $w(t_i, D_k)$	*doctor* 0.3	*hospital* 0.25	*medical* 0.2	*intern* 0.15	*surgery* 0.1
medical	0.9	0.27	0.25	0	0	0.22
football	-0.2	0	0	0	0	0.01
sport	0.7	0	0	0	0	0
nature	0.5	0	0	0	0	0
$\sum_{\forall j/t_j \in T_{u_l}} -1^n \sqrt{w(t_i, D_k)\,\lvert DOI(t_j, D_{u_l})\rvert}$	r_{ij}	0.14	0.12	0	0	0.065
R_{DOI_1}		0.32				

In this manner, $R_{DOI_1}(GA, U) = 0.32$, making the total relationship $R(GA, U) = 0.37$, using $\alpha = 0.5$, which makes the direct relationship as important as the one-hop relationship.

Let us compare this value with the one obtained by calculating the degree of relationship between the user and an episode of the series 'Prison Break' (PB), whose tag cloud is: *prison* (0.45), *runaway* (0.3), *love-story* (0.25). Having a look at the tags in their tag clouds, we should expect that this series is not as recommendable for the user as the previous one. Indeed, our hypothesis is true, since $R_{DOI_0}(PB, U) = 0$ (as they do not share any tag) and $R_{DOI_1}(PB, U) = 0.11$, so that the total relationship is $R_{DOI}(PB, U) = 0.06$.

To measure now the relationship between the user and the course of 'Introduction to Cardiology' we need to calculate both the relationship concerning the DOI and DOK, since it is a learning element. In addition, we need an extra piece of information, the degree of difficulty (DOD) of the learning element. For this example $DOD(IC) = 0$, i.e., *medium*.

First, we calculate the direct relationships, using the only the tag they share: *medical*, resulting $R_{DOI_0}(IC, U) = 0.67$ and $R_{DOK_0}(IC, U) = 0.52$.

The following tables will serve to calculate the one-hop relationships.

	$DOI(t_j, D_{u_l})$ \ $w(t_i, D_k)$	*cardiology* 0.5	*medical* 0.3	*heart* 0.2
medical	0.9	0.3	0	0
football	-0.2	0	0	0
sport	0.7	0	0	0.2
nature	0.5	0	0	0.1
$\sum_{\forall j/t_j \in T_{u_l}} -1^n \sqrt{w(t_i, D_k)\,\lvert DOI(t_j, D_{u_l})\rvert}$	r_{ij}	0.2	0	0.11
R_{DOI_1}		0.31		

Hence, the one-hop relationship concerning the interest of the user is $R_{DOI_1} = 0.31$. For this reason, the total relationship concerning DOI (assuming $\alpha = 0.5$) is $R_{DOI}(IC,U) = 0.49$.

	$w(t_i,D_k)$ \ $DOK(t_j,D_{u_l})$	*cardiology*	*medical*	*heart*
		0.5	0.3	0.2
medical	0.2	0.3	0	0
football	0.5	0	0	0
sport	0.6	0	0	0.2
nature	0.65	0	0	0.1
$\sum_{\forall j/t_j \in T_{u_l}} \sqrt{w(t_i,D_k)\,\frac{1}{2}\,(2-\lvert DOK(t_j,D_{u_l})-DOD(c_k)\rvert)}$	r_{ij}	0.2	0	0.11
R_{DOK_1}		0.31		

Hence, the one-hop relationship taking into account the knowledge of the user is $R_{DOK_1}(IC,U) = 0.31$. For this reason, the total relationship concerning DOK (assuming $\alpha = 0.5$) is $R_{DOK}(IC,U) = 0.42$.

We can see that the course is more appropriate for the user with respect to his/her interests than to his/her knowledge. Calculating the total relationship, assuming that the user assigns the same importance to his/her interests than to his/her knowledge ($\beta = 0.5$), we obtain $R(IC,U) = 0.46$.

If we compare this degree of relationship with the one obtained from comparing the user with the course of 'Labour Risk Prevention' (taking the $DOD = -0.5$, i.e., *easy*) we obtain the results summarised in the following table:

	R_{DOI_0}	R_{DOI_1}	R_{DOI}	R_{DOK_0}	R_{DOK_1}	R_{DOK}	R
Introducción a la Cardiología	0.67	0.31	0.49	0.52	0.31	0.42	0.46
Prevención de Riesgos Laborales	0	0.18	0.09	0	0.2	0.1	0.1

Which shows that the first course is more appropriate for the user than the second one.

5 Stereotyping the Classes of the Ontology

As mentioned in Sec. 2, each piece of content, whether a TV programme or a learning element, is an individual of at least one class of the TV-Anytime ontology or the domain ontology, respectively. Thus, every class of these ontologies is constituted by several contents with their own tag cloud.

In this manner, it seems reasonable that a tag which has a heavy weight in the majority of the tag clouds of the individuals of a class not only describes the individuals themselves but also the future instances of the class, i.e., some other TV programmes of the same genre. For example, the programmes 'Grey's Anatomy', 'ER' and 'Hospital Central' (the Spanish version of 'ER') belong to the class *Medical Drama*. Many users will assign the tags *doctor, hospital* and *medicine* to these programmes. Thus, we can say that these tags not only describe the contents themselves but they have a high probability of describing the future individuals of this class. That is why we create stereotypes for the classes from the tag clouds of their individuals, taking into account the relative index of popularity of the content in the class, i.e., we give more importance to the case that a content that has been tagged by many users has a particular tag in its tag cloud than to the one where the same tag has been assigned to a content described by few users.

First, we calculate the IOP of the content (c_i) regarding to the class it belongs to ($c_i^{\uparrow}$), denoted as $IOP(c_i, c_i^{\uparrow})$ and calculated as follows:

$$IOP(c_i, c_i^{\uparrow}) = \frac{IOP(c_i)}{\sum\limits_{\forall k / c_k \in c_i^{\uparrow}} IOP(c_k)} \tag{21}$$

In this manner, the weight of the tag t_j in the multiset of tags of the stereotype instance of the parent class will be obtained by weighting the weights of the tags by its IOP.

$$w_O(t_j, D_{i_s}) = \sum_{\forall k / c_k \in c_i^{\uparrow}} IOP(c_k, c_i^{\uparrow})\, w(t_j, D_k) \tag{22}$$

This is what we have called *direct stereotyped weight*, since it only takes into account the weights of the tags, not the relationships of these tags with the rest of the tags the other instances have been described with. The total weight of the tags in the stereotype should be calculated as follows:

$$w(t_j, D_{i_s}) = w_O(t_j, D_{i_s}) + w_{REL}(t_j, D_{i_s}) \tag{23}$$

The second term is named *stereotyped weight through relationships* (w_{REL}) and calculated by means of the following equation.

$$w_{REL}(t_j, D_{i_s}) = \sum_{\forall k / c_l \in c_i^{\uparrow}} IOP(c_l, c_i^{\uparrow})\, R_1(t_j, c_l) \tag{24}$$

Being $R_1(t_j, c_l)$ the relationship between the tag t_j and the content c_l, which is calculated using the equation (12), taking into account that c_k is a piece of content whith $T_k = \{t_j\}$.

The last step is normalizing the weights of the tags in the stereotype so that

$$\sum_{\forall k / t_k \in D_{i_S}} \overline{w}(t_k, D_{i_s}) = 1$$

applying the following equation:

$$\overline{w}(t_j, D_{i_s}) = \frac{w(t_j, D_{i_s})}{\sum\limits_{\forall k / t_k \in D_{i_S}} w(t_k, D_{i_s})} \tag{25}$$

Tab. 1 shows an example of how the weight of the tags in the stereotype's tag cloud is calculated from the tag clouds of all the individuals of the class. For example, look at the difference between the tags *nurse* and *romance*. Although the direct stereotyping weight of *romance* is bigger than the *nurse* one, the relationship

Table 1. Propagation of tags

	Weight GA $IOP = 0.45$	Weight ER $IOP = 0.33$	Weight HC $IOP = 0.22$	Direct propagation	Relationship propagation	Total propagation
Operating Room	0.0000	0.1373	0.0000	0.0458	0.9882	0.2682
Intern	0.1538	0.0000	0.0000	0.0684	0.6867	0.1959
Doctor	0.4615	0.3922	0.3279	0.4087	0.1893	0.1551
Hospital	0.3077	0.3922	0.4098	0.3585	0.2157	0.1489
Nurse	0.0000	0.0000	0.0820	0.0182	0.4792	0.1290
Adultery	0.0769	0.0000	0.0000	0.0342	0.1028	0.0355
Romance	0.0000	0.0000	0.1639	0.0364	0.0951	0.0341
Death	0.0000	0.0784	0.0164	0.0298	0.0984	0.0332

stereotyping weight is bigger for *nurse*, since it is more related to the rest of the tags, such as *doctor* or *hospital*, as expected.

This process of creation of stereotypes can be used with several aims. The most direct one is categorizing the contents into a class of the correspondent ontology if this content has not been previously provided with the appropriate metadata. This is accomplished by comparing the content's tag cloud with the tag clouds of the stereotypes of the ontology and determining which one shows a greater similarity, using the mechanism presented in Sec. 4.2. However, the most interesting goal is calculating offline the similarity between the tag clouds of the stereotypes of the concepts with the stereotypes of the TV genres, to determine which are the TV genres that are good to explain particular concepts and vice versa.

The stereotyping method described above serves us to establish relationships between concepts of the domain ontology and genres of TV programmes offline. Thus, when a new TV programme arrives to the set-top box, we can start looking for related learning elements in the classes of the concepts that are highly related with the genre that this TV programme belongs to, making the process shorter.

6 Conclusions and Future Work

In this chapter, we have exposed a new approach to t-learning experiences that allows to create personalised experiences that combine TV programmes and learning elements to attract viewers towards education, using the *entercation* concept that we had previously presented. The novelty of this chapter with respect to previous work is the mechanism that we use to create these experiences, which combines ontologies and folksonomies to calculate the degree of relationship between users and contents and between contents themselves. This mechanism permits the recommendation of contents to the users as well as the identification of related contents to compose the learning experiences. Because of its hybrid approach, it takes the best of both worlds, considering both creators and users' point of view about the contents.

One of the possible problems of the system could be the amount of information provided by the users, which can be huge to be dealt with or to create the folksonomy. However, once the system has achieved a stationary state —i.e. it has overcome the initial period— the users are likely to assign the already existing tags to the new contents, and new tags are very rare. Thus, the size of the folksonomy does not almost increase, although the calculation of the relationships between tags is more accurate if it has been calculated from more programmes' tag clouds.

This proposal has been presented to be used to relate TV programmes and learning objects. However, it is completely applicable to the recommendation and relationship of any contents that are collaboratively tagged and can be considered as individuals of an ontology (if they do not belong to classes of an ontology, the algorithm to calculate the degree of relationship is still applicable but the improvements obtained with ontologies are not). For example, it could be used to offer web pages that are related to the one the user is reading, or related to the video he/she is watching, similar products to the one the user is browsing, etc.

In the future, we intend to improve the algorithm by augmenting the degree of relationship between the contents when the classes they belong to are closely related, as well as determining the concepts or TV genres the users prefers and augmenting the degree of relationship between the content and the user if it belongs to any of these classes.

Another important issue which can be considered is the persistence of tags in the system, for example, in users' tag clouds. In the learning field, it makes sense that the tags are only valid for a limited period of time, e.g. an academic semester. However, in t-learning, learning experiences are more casual and are not structured in time. Nevertheless, it could be interesting to introduce in this proposal the possibility of decreasing the importance of tags in users' tag clouds if they have not been used for a long period of time.

Moreover, we are working on an extension of our proposal that takes into account the fact that TV is a medium that is usually watched in group. In this manner, we should be able to make recommendations to a group of users creating a group tag cloud, so that we can create *entercation* experiences that are appropriate for the whole group.

Last but not least, there are some other fields which can benefit from this proposal, for example, m-learning, using IDTV in mobile devices. In this paradigm, the personalisation of experiences is even more important since users are supposed to use these devices for learning in short spare periods of time. For this reason, the approach exposed in this paper can be translated to this medium but the granularity should be refined, i.e. the pieces of content that are offered should be shorter, such as pieces of TV programmes. Hence, we need to establish a mechanism which allows to add tags to segments of contents instead of full ones.

Acknowledgements. Funded by the Ministerio de Educación y Ciencia research project TSI2007-61599, by the Consellería de Educación e Ordenación Universitaria incentives file 2007/000016-0, and by the Programa de Promoción Xeral da Investigación de Consellería de Innovación, Industria e Comercio research project PGIDIT05PXIC32204PN.

References

1. Advanced Distributed Learning (ADL): Sharable Content Object Reference Model (SCORM ®) 2004, 3rd edn. (2006), http://www.adlnet.org
2. Advanced Distributed Learning (ADL): Sharable Content Object Reference Model (SCORM®) 2004, 3rd edn. Run-Time Environment Version 1.0. (2006), http://www.cen-ltso.net/
3. Golder, S.A., Huberman, B.A.: The structure of collaborative tagging systems. Journal of Information Science 32(2), 198–208 (2006)
4. Hotho, A., Jäschke, R., Schmitz, C., Stumme, G.: Information Retrieval in Folksonomies: Search and Ranking. In: Sure, Y., Domingue, J. (eds.) ESWC 2006. LNCS, vol. 4011, pp. 411–426. Springer, Heidelberg (2006)
5. IEEE Learning Technology Standards Committee (LTSC): Learning Object Metadata. IEEE Standard 1484.12.1 (2002)
6. Michlmayr, E., Cayzer, S., Shabajee, P.: Add-A-Tag: Learning Adaptive User Profiles from Bookmark Collections. In: 1st International Conference on Weblogs and Social Media (ICWSM 2007), Boulder, USA, Colorado (2007)
7. Niwa, S., Doi, T., Honiden, S.: Web Page Recommender System based on Folksonomy Mining for ITNG 2006 Submissions. In: Society, I.C. (ed.) Third International Conference on Information Technology: New Generations (ITNG 2006), Washington, DC, USA, pp. 388–393 (2006)
8. Rey-López, M., Fernández-Vilas, A., Díaz-Redondo, R.P.: New trends for personalised t-learning. In: Arias, J.J.P. (ed.) Personalization of Interactive Multimedia Services: A Research and Development Perspective, pp. 149–162. Nova Science Publishers, Inc., Bombay (2008)
9. Rey-López, M., Díaz-Redondo, R.P., Fernández-Vilas, A., Pazos-Arias, J.J.: Entercation: Engaging Viewers in Education through TV. ACM Computers in Entertainment 5(2) (2007)
10. Specia, L., Motta, E.: Integrating folksonomies with the semantic web. In: The Semantic Web: Research and Applications, pp. 624–639 (2007)
11. Szomszor, M., Cattuto, C., Alani, H., O'Hara, K., Baldassarri, A., Loreto, V., Servedio, V.D.: Folksonomies, the Semantic Web, and Movie Recommendation. In: 4th European Semantic Web Conference, Bridging the Gap between Semantic Web and Web 2.0, Innsbruck, Austria, pp. 71–84 (2007)
12. The TV-Anytime Forum: Broadcast and On-line Services: Search, select and rightful use of content on personal storage systems. European Standard ETSI TS 102 822 (2004)
13. Vander Wal, T.: Folksonomy Coinage and Definition (2007), http://www.vanderwal.net/folksonomy.html

Computational Intelligence Infrastructure in Support for Complex e-Learning Systems

Santi Caballé[1] and Fatos Xhafa[2]

[1] Department of Computer Science, Multimedia and Telecommunication
Open University of Catalonia,
Rambla Poblenou, 156, 08018 Barcelona, Spain
[2] Department of Languages and Informatics Systems
Technical University of Catalonia
Jordi Girona, 1-3, 08034 Barcelona, Spain

Abstract. Modern on-line learning environments are to enable and scale the involvement of an increasing large number of single/group participants who can geographically be distributed, and who need to transparently share a huge variety of both software and hardware distributed learning resources. As a result, collaborative learning applications are to overcome important non-functional requirements arisen in distributed environments, such as scalability, flexibility, availability, interoperability, and integration of different, heterogeneous, and legacy collaborative learning systems. Therefore, e-Learning applications need to be developed in a way that overcome these demanding requirements as well as provide educational organizations with fast, flexible and effective solutions for the enhancement and improvement of the learning performance and outcomes. In this chapter, an innovative engineering software technique is presented and validated that combines the Generic Programming paradigm and Service-Oriented Architectures for the effective and timely construction of flexible, distributed, scalable, interoperable and robust applications as key aspects to address the current demanding and changing requirements in the current e-Learning domain.

1 Introduction

Over the last decade, educational organizations' needs have been changing in accordance with ever more complex pedagogical models as well as with technological evolution resulting in e-Learning environments with very dynamic and changing teaching and learning requirements, especially for collaborative learning [22], [11].

In particular, these needs involve extending and moving to highly customized learning and teaching forms in timely fashion, each incorporating its own pedagogical approach, each targeting a specific learning goal, and each incorporating its specific resources. Organizations' demands also include a cost-effective integration of legacy and separated learning systems, from different institutions, departments and courses, which are implemented in different languages, supported by heterogeneous platforms and distributed everywhere, to name some of them [25].

F. Xhafa et al. (Eds.): Computational Intelligence for Tech. Enhanced Learning, SCI 273, pp. 143–168.
springerlink.com

From this view, one of the main challenges in the development of modern e-Learning systems is to overcome important requirements arisen in distributed environments such as scalability, flexibility, availability, mobility, interoperability, and integration of different, heterogeneous, and legacy learning systems. In this chapter, we focus on certain non-functional requirements which are especially frustrating when they are not fulfilled appropriately during the online learning activity, such as fault-tolerance, scalability, performance, and interoperability. They may have considerable repercussions on the learning performance and outcomes as their lack impedes the normal learning flow as well as discriminates learners in terms of technology skills and technical equipment. In particular [14].

- Fault-tolerance refers to provide permanent access to the learning environment and its resources from anywhere and at any time, even in the presence of system failures. The temporary lack of service due to technical difficulties prevents students from fulfilling the learning task in time as well as to meet and collaborate with others as scheduled. Hence, it is expected that the learning system is able to seamlessly recover from failures and keep providing the service.
- A system is told to be scalable if it is able to bear a growing load of both resources and users who may access concurrently to these resources as well as to offer data-intensive and complex functionalities without causing performance repercussions on the underlying system supporting the learning environment. Indeed, a system performing poorly is one of the most frustrating aspects during the on-line collaborative learning experience as it makes participants' requests be waiting for long periods of time to be served. In order to keep on providing a high level of quality of service, a learning system should seamlessly scale to new resources of both hardware and soft-ware at the same pace as the workload increases.
- The gain in performance might help, for instance, include more complex information of the collaboration to be generated and presented in real time (such as modeling the participants' behavior during the discussion by combining individual and group session and navigation information) as well as to make an in-depth analysis through data mining techniques to provide tutors with ongoing progress of students learning during the discussion activity.
- The benefits from having an interoperable environment is to provide students and tutors with transparent access and share of a large variety of both software and hardware learning resources. Users using any computer platform and software should be able to collaborate among them and interoperate with all the existing resources. Interoperability includes supporting users with little technology skills by avoiding the need to set up anything on the client side. Moreover, this may improve the overall learning experience by avoiding the repercussions derived from redundancy and inconsistency of existing databases and information systems in general (e.g., unifying the authentication process so that the user has access to all learning tools by logging in them just once) and integrating seamlessly external and legacy tools and applications.

These entire requirements represent a great challenge for the latest trends of software development to be completely satisfied. To this end, Service-oriented

Architectures (SOA) [31] has emerged to boost educational organizations by increasing the flexibility of their pedagogical strategies, which can be continuously adapted, adjusted, and personalized to each specific target learning group. Moreover, SOA facilitates the reutilisation of successful collaborative learning experiences and makes it possible for the online learning participants to easily adapt and integrate their current best practices and existing well-known e-learning tools into new learning goals.

Moreover, in order to face the current and costly complex e-learning needs, educational organizations and software companies leverage the benefits from software reusability [15], in terms of quality, productivity and cost. Indeed, the application of reused-based techniques, such as Generic Programming (GP) [9], appear to be a good choice for the systematic and timely development of complex and high quality e-learning applications.

Therefore, software techniques and paradigms have been evolving all the time [12], [15] to mainly provide higher levels of abstraction and transparency so that developers can reuse and integrate not only functionality and components but more complex yet larger pieces of software. Indeed, transparency plays a key role in the design of appropriate learning systems that support strong fault-tolerance, dynamic scalability and inherent interoperability, among others requirements. However, although transparency has been greatly enhanced in the e-Learning domain, the barrier of technology incompatibilities and a poor quality of service make the learners' experience ineffective and difficult when interacting with the learning system [5].

To sum up, service-oriented architectures have come to play a major role in the context of e-Learning [18] due to the benefits that provide in terms of interoperability among heterogeneous hardware and software platforms, integration of new and legacy systems, flexibility in updating software, and so on. In addition, distributed technologies, such as Grid, have emerged to extend to a large-scale, flexible, secure, coordinated resource sharing among dynamic collections of individuals, institutions, and resources [14]. These features form an ideal context for supporting and meeting the mentioned demanding requirements of collaborative learning applications.

In this chapter, these entire approaches are taken further through the intensive use of the Collaborative Learning Purpose Library (CLPL) [6] based on reused-based fine-grained Web-services [31], especially designed to take great advantage of service-oriented and distributed technology and help develop enhanced collaborative learning systems. The ultimate goal is to provide support for meeting the demanding requirements found in the e-learning domain and specifically improve the effectiveness of the collaborative learning experience. To this end, first we provide a complete background of the main concepts and technologies used in this chapter. Form this background, we present the aims and principles of the CLPL. Then, an e-Learning application to validate the CLPL is presented using distributed infrastructure and finally the chapter shows the experimental results achieved. The chapter closes by summarizing the key points of the approach presented as well as outlining ongoing and future work.

2 Background

In this section, a brief overview of the existing technologies and paradigms related to this work is presented, namely Computer-Supported Collaborative Learning, Generic Programming, Service-Oriented Architecture, Model-Driven Architecture, and Distributed Computing. This overview will serve as background for the next sections.

2.1 Computer-Supported Collaborative Learning

Computer-Supported Collaborative Learning (CSCL) is one of the most influencing research paradigms dedicated to improve teaching and learning with the help of modern information and communication technology [22], [11], [30], [29], [10]. Collaborative or group learning refers to instructional methods where students are encouraged to work together on learning tasks. As an example, project-based collaborative learning proves to be a very successful method to that end [11]. Therefore, CSCL applications aim to create virtual collaborative learning environments where students, teachers, tutors, etc., are able to cooperate with each other in order to accomplish a common learning goal.

To achieve this goal, CSCL applications provide support to three essential aspects of collaboration, namely coordination, collaboration and communication; with communication being the base for reaching coordination and collaboration [26]. Collaboration and communication might be synchronous or asynchronous. The former means cooperation at the same time and the shared resource will not typically have a lifespan beyond the sharing while the latter means cooperation at different times being the shared resource stored in a persistent support.

The representation and analysis of group activity interaction is an important issue in CSCL. Data analysis of the information captured from the actions performed by learners is indeed a core function for support of coaching and evaluation in online collaborative learning environments as well as the modeling of the learner's behavior during the learning process and of the learning process itself [11], [20], [28], [27], [10]. To this end, fine-grained notifications and complex information collected from the learners' interaction are provided to give immediate feedback about others' activities and about the collaboration in general [32].

2.2 Generic Programming

In all advanced forms of engineering it can be observed that new products are usually developed by reusing tried and tested parts rather than developing them from scratch. The reuse of previously created product parts leads to reduced costs and improved productivity and quality to such an extent that industrial processes will take a great leap forward. Generic Programming (GP) [9] has emerged over the last years to facilitate this possibility in the software engineering field.

GP is an innovative paradigm that attempts to make software as general as possible without losing efficiency. It achieves its goal by identifying interrelated high-level family from a common requirement set. By the application of this technique, especially in design phases, software is developed offering a high degree of abstraction which is applicable to a wide range of situations and domains.

By applying GP to develop computer software important objectives are achieved [8]:

- Reuse. This means to be able to reuse and extend software components widely so that it adapts to a great number of interrelated problems.
- Quality. Here "quality" refers to the correctness and robustness of implementation which provides the required degree of reliability.
- Efficiency. It is also essential to guarantee the efficiency of components as if this not done the performance repercussions will be noted, just as with lack of quality, in all of the systems involved.
- Productivity. Inherent to reutilization is the saving through not having to create software components again that already exist. Hence, there is an increase in computing production.
- Automation. The aim is to automate the processes so that general requirements with a high level of abstraction and specially designed tools can be used to produce operative programmes.
- Personalisation. As the general requirements are made more particular, so the product that is generated becomes more optimised to meet the specific needs of the client.

GP also represents one important technique to achieve effective Product Lines (PL) following the Product-Line Architecture (PLA) approach [15]. PLA promotes developing large families of related software applications quickly and cheaply from reusable components. In PLA, a certain level of automation is provided in the form of generators (also known as component configuration tools) to realize solutions for large parts of the systems being developed.

2.3 *Service-Oriented Architecture*

Service-Oriented Architecture (SOA) [31] represents the next step in the software development to help organizations meet their ever more complex set of needs and challenges, especially in distributed systems [16], [6]. This is achieved by dynamically discovering and invoking the appropriate services to perform a request from heterogeneous environments, regardless of the details and differences of these environments. By making the service independent from the context, SOA provides software with important non-functional capabilities for distributed environments (such as scalability, heterogeneity and openness), and makes the integration processes much easier to achieve.

SOA relies on services. According to W3C [31], a service is a set of actions that form a coherent whole from the point of view of service providers and service requesters. In other words, services represent the behaviour provided by a provider and used by any requesters based only on the interface contract. Within SOA, services.

- stress location transparency by allowing services to be implemented, replicated and moved to other machines without the requester's knowledge,
- enable dynamic access as services are located, bound and invoked at runtime,
- promote interoperability making it possible for different organisations supported by heterogeneous hardware and software platforms to share and use the same services,
- facilitate integration of other existing systems and thus protect previous investments (e.g. legacy assets),
- rely on encapsulation as they are independent from other services and their context,
- enhance flexibility by allowing services to be replaced without causing repercussions on the underlying systems involved,
- foster composition from other finer-grained services.

Although SOA can be realised with other technologies, over the last few years Web services has come to play a major role in SOA due to lower costs of integration along with flexibility and simplification of configuration. The core structure of Web services is formed by a set of widely adopted protocols and standards, such as XML, SOAP, WSDL, and UDDI [31], which provide a suitable technology to implement the key requirements of SOA. This is so because these protocols allow a service to be platform - and language - independent, dynamically located and invoked, interoperable over different organization networks, and supported by large organisations (e.g., W3C consortium).

2.4 Model-Driven Architecture

The Model-Driven Development (MDD) paradigm and the framework supporting it, namely Model-Driven Architecture (MDA) [24] have been recently attracting a lot of attention given that it allows software developers and organizations to capture every important aspect of a software system through appropriate models [15]. MDA provides great advantages in terms of complete support to the whole cycle development, cost reduction, software quality, reusability, independence from the technology, integration with existing systems, scalability and robustness, flexible evolution of software and standardization, as it is supported by the Object Management Group [24].

In proposing MDA, two key ideas have had significant influence in OMG aiming at addressing the current challenges in software development [4]: service-oriented architectures (SOA) and product line architectures (PLA). As to the former, SOA provides great flexibility to system architectures by organizing the system as a collection of encapsulated services. Hence, SOA relies on services which represent the behavior provided by a component to be met and used by any other components based only on the interface contract. As to the latter, PLA promotes developing large families of related software applications quickly and cheaply from reusable components.

There are many views and opinions about what MDA is and is not. However, the OMG, as the most authoritative view, focuses MDA on a central vision [24]:

Allow developers to express applications independently of specific implementation platforms (such as a given programming language or middleware). To this end, OMG proposes the following principles for MDA developments: first, the development of a UML-based Platform Independent Model (PIM), second, one or several models which are Platform Specific Models (PSM). Finally, a certain degree of automation by means of descriptions is necessary for mapping from PIM to PSM.

2.5 Distributed Computing

A computational software may greatly enhance its possibilities by taking advantage of the inherent performance potential of distributed technologies, such as Grid and peer-to-peer (P2P) [2]. Furthermore, a set of interesting non-functional features may be provided by these distributed technologies, which represent an ideal context for supporting and producing major benefits for applications. Such important features include [25], [14]: enable and scale the involvement of an increasing large number of single/group participants who can be widely separated by geography and/or time, possibly situated at very different locations, and transparently share a huge variety of both software and hardware resources, even in real time.

The combination of distributed technologies with service-oriented architectures (SOA) allows developers to cope with essential issues, such as integration, interoperability reliability, and flexibility so as to meet the needs of different, heterogeneous and legacy environments [17]. These non-functional requirements found in many domains are to be addressed by software computational platforms in a transparent manner by leveraging the latest software development methodologies, and especially the service-oriented approach. The ultimate goal of computational platforms is to serve as mediators between the application layer and the infrastructure layer making applications independent from the type and evolution of the underlying infrastructure.

Further demanding non-functional requirements found in collaborative learning include the provision of effective information and knowledge about what is occurring in the group activity. This implies receiving knowledge simultaneously both synchronously and asynchronously since the current and history interaction data shown are continuously updated. However, these applications are characterized by a high degree of user-user interaction and hence generate a huge amount of information. This information may also include a great variety of types and formats and hence tends to be large in size [7]. Therefore, the supply of efficient and transparent knowledge to users in this context is a significant challenge. Recently, Grid technology is also increasingly being used in this context to reduce the overall, censored time in processing data by taking advantage of its large computing support and thus facilitating the group activity, decision making, task accomplishment, and assessment of the progress of the group etc.

Furthermore, Grid technology provides a set of interesting features which represent an ideal context for supporting and producing major benefits for collaborative learning applications [1], [14]. Such important features include: large scale of

Grid infrastructures, wide geographical distribution of resources, multiple administrations from different organizations, transparent and dependable access as well as the capability of granting access to shared, heterogeneous resources in very dynamic environments [2]. Considering the benefits provided by distributed technologies it is possible for educational organizations to make use of true collaborative learning environments that enable the involvement of large number of single/group participants (teachers, students, tutors among others) who can potentially belong to many different organizations, possibly situated at very different locations, and transparently share a huge variety of both software and hardware resources while enhancing human-to-human interaction (synchronously or asynchronously) through a friendly complex user interface.

To sum up, leveraging the inherent performance potential of distributed infrastructure for collaborative learning applications makes it possible to greatly enhance the collaboration between users in terms of both participant scalability (adding as many participants/groups as necessary) and resource availability (replicating and executing them in multiple nodes) enabling collaboration as the most important learning method. In addition, distributed infrastructure may help overcome important barriers in the form of certain non-functional requirements, such as fault-tolerance, performance, and interoperability, which are especially frustrating when they are not fulfilled appropriately during the collaborative learning activity. Indeed, these requirements may have considerable repercussions on the learning performance and outcomes as their lack impedes the normal learning flow as well as discriminates learners in terms of technology skills and technical equipment [5].

3 A Service-Oriented Collaborative Learning Platform for Distributed Environments

In this section, it is presented a generic, robust, interoperable, reusable, component-based and service-oriented platform called Collaborative Learning Purpose Library (CLPL) [6]. The main guidelines that conducted its design are shown that allows CSCL applications to take great advantage of distributed infrastructure. The ultimate goal of the CLPL is to provide support for meeting the demanding requirements found in the CSCL domain and considerably improve the effectiveness of the collaborative learning experience.

A computational software may greatly enhance its possibilities by taking advantage of the inherent performance potential of distributed technologies, such as Grid and peer-to-peer (P2P) [2]. Furthermore, a set of interesting non-functional.

3.1 The Design and Implementation of the CLPL

There a great deal of similarities between the pervasive and challenging collaborative learning needs and the benefits provided by SOA. As a result of this matching, SOA appears to be the best choice to support the development of the CLPL. Indeed, SOA enhances educational organizations by increasing the flexibility of their pedagogical strategies, which can be continuously adapted, adjusted, and personalized to each specific target learning group. Moreover, SOA facilitates the reutilisation of successful collaborative learning experiences and makes it possible for the collaborative learning participants to easily adapt and integrate their current best practices and existing well-known learning tools into new learning goals.

Over the last years, collaborative learning has become a complex and extensive domain. Therefore, the application of the GP principles appear to be a good choice for the development of the CLPL by, first, identifying those parts which are common to most applications of the collaborative learning domain. Then, proceed to isolate the fundamental parts in the form of abstractions from which the basic requirements are obtained. Finally, encourage the greatest possible reusability of the resulting generic components for the construction of as many collaborative learning applications as possible.

In order to turn the CLPL into an effective software platform, its development was based on the MDA approach for the e-Learning domain [23]. This paradigm fits very well in this context in combination with the GP and SOA principles due to the clear separation of a generic, reusable technology-independent model from a different, flexible technology-dependent implementation models.

To this end, the first step was to create a PIM by applying the following Generic Programming ideas [8]: (i) define the semantics of the properties and domain concepts, (ii) extract and specify the common and variable properties and their dependencies in the form of abstractions found in the CSCL domain, and (iii) isolate the fundamental parts in the form of abstractions from which the basic requirements were obtained, analysed and designed as a traditional three-layer architecture (i.e. presentation, business and information). To this end, the PIM was expressed using UML as the standard modelling language promoted by the OMG.

The second step was to build two different PSM from the unique PIM achieved: A Java implementation in the form of a generic component-based library and a collection of WSDL files organized in directories that are automatically turned into generic web-services implemented in the desired programming language and allowing developers to implement the services according to specific needs. On the one hand, the Java programming language provides great predisposition to the adaptation and correct transmission of generic software design, which make the software highly reusable. It lacks of full interoperability between different programming languages though. On the other hand, in order to increase flexibility and interoperability, the SOA-based PSM provides great predisposition to be involved in distributed environments supporting different middleware and programming languages.

Finally, in order to automate as much as possible the transition from the PIM to the appropriate PSM, the latest research results are leading us to deal with XMI files (see [24] for details), which are XML-tagged files as the result of coding UML diagrams. In combination with XSL style sheets, it is possible to turn the PIM's XMI files into WSDL files, which represent the input for a Web-service working environment to transform them into a specific-language architecture design (PSM). Lack of comply with standard of the existing UML case tools is the major problem to face next as well as how to provide a more complete and detailed realization of the desired PSM.

The development of the CLPL fully followed the first and second steps while ongoing work is dealing with the last by introducing certain level of automation by means of WSDL descriptions.

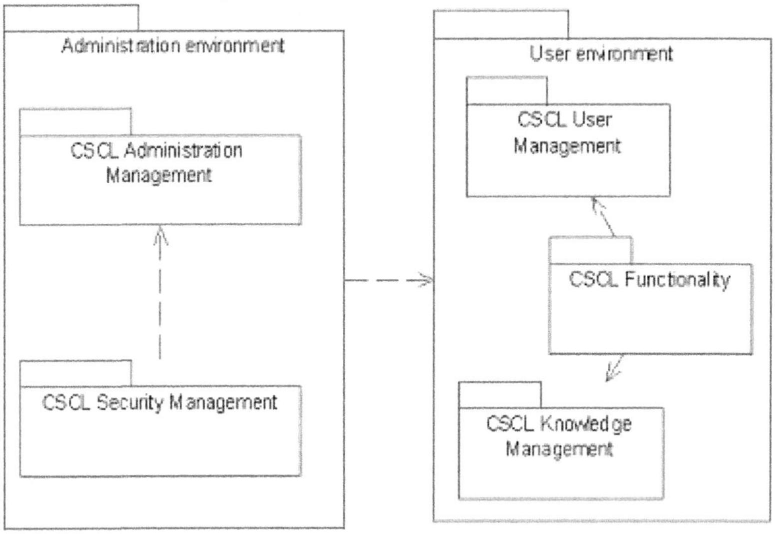

Fig. 1. Graphical representation of the components of the CLPL platform and their dependencies

The CLPL is mainly made up of five components which are independent according to its general internal functionality (see complete description in [6] and also Figure 1:

- CSCL User Management component: this contains all the logics related to the CSCL system user management which can act as a group coordinator, group member, group-entity and system administrator. It tackles both the basic user management functions in a learning environment and the user profile management. The latter implements the user and group models within a collaborative environment.
- CSCL Security Management component: this contains all the generic descriptions of the measures and rules decided to carry out the authentication and authorization issues and so protecting the system from both the unknown users and the intentional or accidental bad use of its resources. Its genericity lets programmer implement them with the ultimate cryptographic security mechanism existing.
- CSCL Administration Management component: this contains those specific data (through log files) and processes (statistical computations) so as to carry out all system's control and maintenance with the aim to administer the system correctly and to improve it in terms of performance and security.

- CSCL Knowledge Management component: This supports a complete process of information and knowledge management. First, this component collects and classifies the system log files made up of all the events occurring in a certain workspace over a given period of time. Then, it performs the statistical analysis on the event information as well as the management and maintenance of the knowledge generated from this analysis.
- CSCL Functionality component: this defines the three elemental parts involved in any form of cooperation, namely coordination, communication and collaboration [13]. Coordination involves the organization of groups to accomplish the important objectives of members such as workspace organization and group structure and planning. Collaboration lets group members share any kind of resources while communication represents the basis of the whole component since it enables coordination and collaboration to be achieved by providing them with low-level communication support. Furthermore, this component implements the presentation to users of the knowledge extracted by the previous component in terms of immediate awareness and constant feedback of what is going on in the system.

3.2 The CLPL on a Distributed Infrastructure

In order to fulfil the functionalities designed in the CLPL, the primary principle was to provide a broad set of independent fine-grained services grouped by a particular purpose, such as the authentication process and the presentation of the feedback extracted. The goal was both to enhance the flexibility in the development of CSCL applications and to ease the deployment of these applications in a distributed context.

To this end, each particular behavior of the CLPL is discomposed into three specialized Web-services matching each of the three layers of a typical software development, namely user interface, business and data [15]. As a result, the completeness of each specific behavior goes through three separate, necessary, sequential steps that connect to the client on one side and to the persistent storage (e.g., database) on the other side. For instance, the authentication process is formed by three different, independent Web-services, namely the authentication user interface, the authentication business, and the authentication data. Thus, when the user attempts to log in, the client code calls the authentication user interface Web-service, which is in charge of collecting the credentials presented by the user. Then, this Web-service calls in turn the authentication business Web-service so as to verify the correctness of the user's input (e.g., input no blank, well-formatted, etc.). Moreover, as part of the business process, this Web-service validates the users' input upon the information existing in the data-base by calling the authentication data Web-service, which is responsible for accessing the database and extracting the authentication data of the user.

A clear, independent, and separated vision of each single behavior of the CLPL into fine-grained task-specific Web-services results in a natural distribution of the application into different nodes in a network. This distribution is driven by matching each Web-service's purpose to the most appropriate node's configuration and location in the network. According to this view, the Web-services in the user interface layer should be allocated nearby the client; the business Web-services would be better suited if allocated in those nodes with high-performance processors, and, finally, the data Web-services could be attached or nearby the database supported by nodes with high storage capability. As for the database, it can also be distributed as it is clearly separated from the data Web-services, which would be in charge of updating and keeping the consistency of the different instances of the database.

The work methodology proposed by the CLPL offers throughout flexibility as to where (i.e., network node) to install both each learning system function (i.e., CSCL behavior) and each layer of this function (i.e., Web-service). Moreover, the widely adopted standards of the Web-services technology (e.g., HTTP and TCP/IP [31] help communicate the Web-services with each other in a network just using their IP ad-dress and passing through firewalls and other barriers that other technologies have problems to overcome. On the other hand, there exist many open-source technologies that deal with Web-services, such as Apache Tomcat and Axis, allowing developers to easily use and deploy the services provided by the CLPL.

In this context, both the independence between the fine-grained services provided by the CLPL and the use of key techniques found in the typical distributed development, such as replication, produce many important benefits. Indeed, by installing and deploying replicas of the Web-services all over the network fault-tolerance is easily achieved by redirecting a request to an exact replica of the Web-service when a node is down. Concurrency and scalability become natural in this context by parallelizing the users' requests using as many replicas as necessary. Furthermore, load balancing can be achieved so as to increase the overall performance of the system. Finally, inter-operability is inherent in the context of Web-services technology as they are fully independent from software and hardware platforms and programming languages.

To sum up, combining the generic view of CSCL domain provided by the CLPL, the Web-services technology, and leveraging distributed infrastructure, the realization of the most demanding requirements arisen in CSCL environments becomes a reality.

4 An Application Example: A Distributed Discussion Forum

To illustrate the entire approach, a Web-based structured discussion forum called Discussion Forum (DF) was developed to validate the possibilities offered by the CLPL and the use of distributed infrastructure to enhance the collaborative learning experience. The ultimate aim was to demonstrate both the effectiveness of developing a complex system in terms of quality, productivity and cost and the provision of new opportunities to learning methodologies, such as learning by discussion, that gives significant benefits to students in the context of project-based learning, and in education in general. This applications is currently running at the Open University of Catalonia providing support to in-class discussion assignments in several on-line courses.

In this section, first the pedagogical requirements of the DF are described and then some guidelines that conducted its design are provided. Finally, the implementation and deployment of the DF is reported.

4.1 Pedagogical Background

In collaborative learning environments, the discussion process forms an important social task where participants can think about the activity being performed, collaborate with each other through the exchange of ideas that may arise, propose new resolution mechanisms, and justify and refine their own contributions and thus acquire new knowledge [21].

To this end, a complete discussion and reasoning process is proposed based on three types of generic contributions, namely specification, elaboration and consensus [29]. Specification occurs during the initial stage of the process carried out by the tutor or group coordinator who contributes by defining the group activity and its objectives (i.e. statement of the problem) and the way to structure the group activity in sub-activities. Elaboration refers to the contributions of participants (mostly students) in which a proposal, idea or plan to reach a solution is presented. The other participants can elaborate on this proposal through different types of participation such as questions, comments, explanations and agree/disagree statements. Finally, when a correct proposal of solution is achieved, the consensus contributions take part for its approval (this includes different consensus models such as voting); when a solution is accepted the discussion terminates.

4.2 The Design and Implementation of the Application

The design of the DF includes certain thematic annotation cards (such as idea, evaluation, reply) that structure the elaboration phase and can offer full help support as well (see Figure 2). All events generated are to be recorded as user actions, which are then analyzed and presented as information to participants either in real time (to guide directly students during the learning activity) or after the task is over (in order to understand the collaborative process).

To that end, the CSCL Knowledge Management and CSCL Functionality components provided full support to the event management. In particular, the design of the DF includes certain thematic annotation tags based on the low-level exchange categories (see [3] for a complete description), such as information-clarification and request of opinion, which qualifies each contribution and as a result structure the discussion process (Figure 2). In order to avoid unnecessary choice, each context of the discussion process determines a precise and short list of just those categories that are possible in a certain point of the discussion process (e.g., in replying any kind of request, just the cards involving the provision of information are provided to classify the reply). This makes the choice of the appropriate tag much shorter and easier and no error-prone. In addition, as part of the design, the tutor is to examine and assess all contributions based partially on the tags used by students to categorize them, and as a result students are aware of the potential repercussions of tagging posts incorrectly in order to optimize the assessment instead of reflecting the true meaning of their posts. Equally, group analysis outcomes produced by the treatment of group functioning events constitute an important data source that can assist in achieving a more satisfactory solution to the problem during the consensus phase. Furthermore, the coordinator can use this same information to organize well-balanced groups during the specification phase.

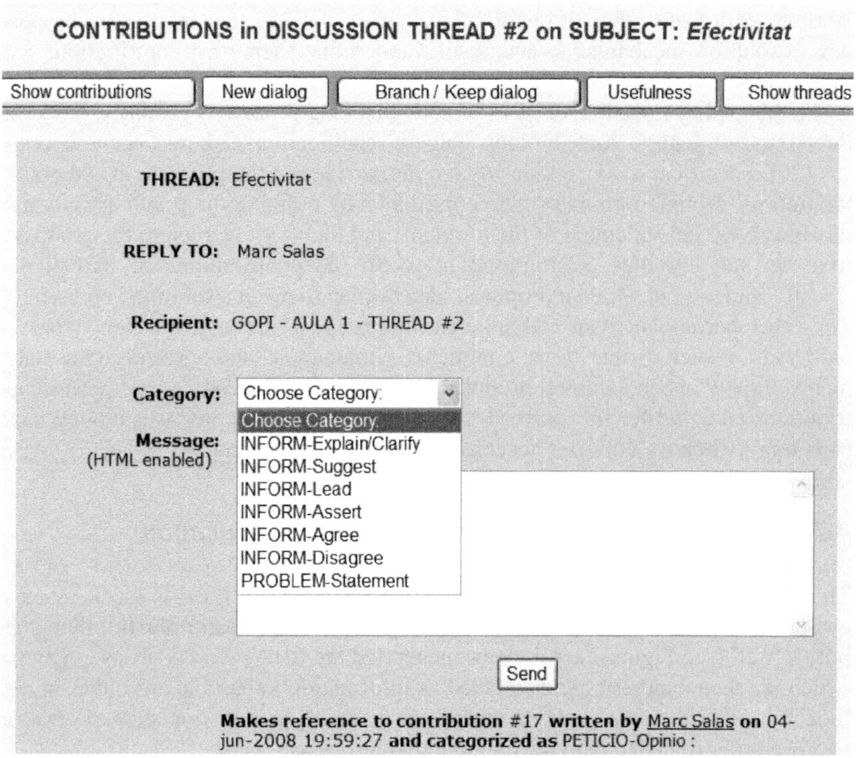

Fig. 2. A specific list of cards for a reply to a contribution categorized as INFORM-Explain

As a result, all contributions are recorded as exchange acts, analyzed and presented as information to participants either in real time (to guide directly students during the learning activity) or after the task is over (in order to understand the collaborative process). To this end, the CLPL's Knowledge Management component provided full support to the interaction management. In particular, a complete treatment of the structured interaction generated enabled the system to keep participants aware of the contributing behavior and dynamics of others, to check certain argumentative structures during discussion and assist in achieving a more satisfactory solution to the problem during the consensus phase, and finally to provide feedback based on the data produced.

FOLDER: #3 - Debat 2

Description: Espai de debat 2.

Created by: Marius Gomez [TUTOR] **on** 25-may-2008 12:00:03

FOLDER DATA:

N. contributions:: 90 **Quality of this folder:** C+ **Usefulness of this folder:** 6.1/10 [227 Vots]

STUDENT STATISTICS

STUDENT POSITION		ACTIVITY				PASSIVITY		IMPACT		EFFECTIVITY	ASSESSMENT		
Pos.	Student	Total contributions	Proactivity	Reactivity	Support	Pending to read	Pending to evaluate	Particip. impact	Replies received	Assientment rebut	Peer assessment	Tutor assessment	FINAL MARK
[1]	Andreu Cuartiella	2/90 2.2%	1/2 50%	1/2 50%	0/2 0%	85/88	86/88	-0.5	1/2	5/5 100%	6.2/10 (4)	B	6.58
[2]	Francisco Garcia	11/90 12.2%	2/11 18%	9/11 81%	0/11 0%	0/79	0/79	-4.5	4/11	3/3 100%	6.3/10 (12)	B	6.57
[3]	JuanPablo Nieto	4/90 4.4%	4/4 100%	0/4 0%	0/4 0%	51/86	86/86	16.5	5/4	17/18 94%	6.8/10 (17)	B	6.53
[4]	Alejandro Lluvia	5/90 5.6%	2/5 40%	3/5 60%	0/5 0%	23/85	42/85	4.5	3/5	7/9 77%	6.7/10 (11)	B	6.46
[5]	Joan Barcelo	4/90 4.4%	2/4 50%	2/4 50%	0/4 0%	39/86	39/86	7.0	4/4	15/16 93%	6.0/10 (9)	C+	6.21
[6]	Miquel Ollers	7/90 7.8%	1/7 14%	6/7 85%	0/7 0%	52/83	67/83	6.5	8/7	42/48 87%	6.4/10 (23)	B	6.15
[7]	Manel Herrera	2/90 2.2%	1/2 50%	1/2 50%	0/2 0%	54/88	61/88	9.5	0/2	12/12 100%	6.6/10 (5)	C+	6.14
[8]	Vanesa Pose	2/90 2.2%	0/2 0%	2/2 100%	0/2 0%	85/88	88/88	-1.0	1/2	17/18 94%	6.7/10 (7)	B	6.05
[9]	Maria Guiu	5/90 5.6%	0/5 0%	5/5 100%	0/5 0%	40/85	85/85	-2.5	4/5	15/18 83%	6.1/10 (17)	B	5.84
[10]	Laura Risco	6/90 6.7%	1/6 16%	5/6 83%	0/6 0%	67/84	84/84	-2.5	5/6	16/17 94%	5.4/10 (15)	C+	5.8
[11]	Francisco Rios	12/90 13.3%	1/12 8%	11/12 91%	0/12 0%	0/78	78/78	1.5	10/12	38/46 82%	6.3/10 (27)	C+	5.75
[12]	Pau Ubach	10/90 11.1%	1/10 10%	9/10 90%	0/10 0%	54/80	66/80	-4.5	7/10	20/23 87%	5.4/10 (18)	C+	5.59
[13]	MariaTeresa Mestre	1/90 1.1%	1/1 100%	0/1 0%	0/1 0%	87/89	89/89	9.0	0/1	9/9 100%	5.2/10 (5)	C+	5.47
[14]	SergioJose Mallorquin	4/90 4.4%	2/4 50%	2/4 50%	0/4 0%	37/86	85/86	5.0	2/4	7/9 77%	6.2/10 (12)	C+	5.33
[15]	Pere Urbon	3/90 3.3%	2/3 66%	1/3 33%	0/3 0%	80/87	84/87	2.5	3/3	4/6 66%	5.9/10 (13)	C+	5.29
[16]	Pere Madrona	2/90 2.2%	0/2 0%	2/2 100%	0/2 0%	68/88	88/88	-1.0	0/2	0/0 0%	5.8/10 (5)	B	4.53

Fig. 3. Monitoring information provided to the tutor

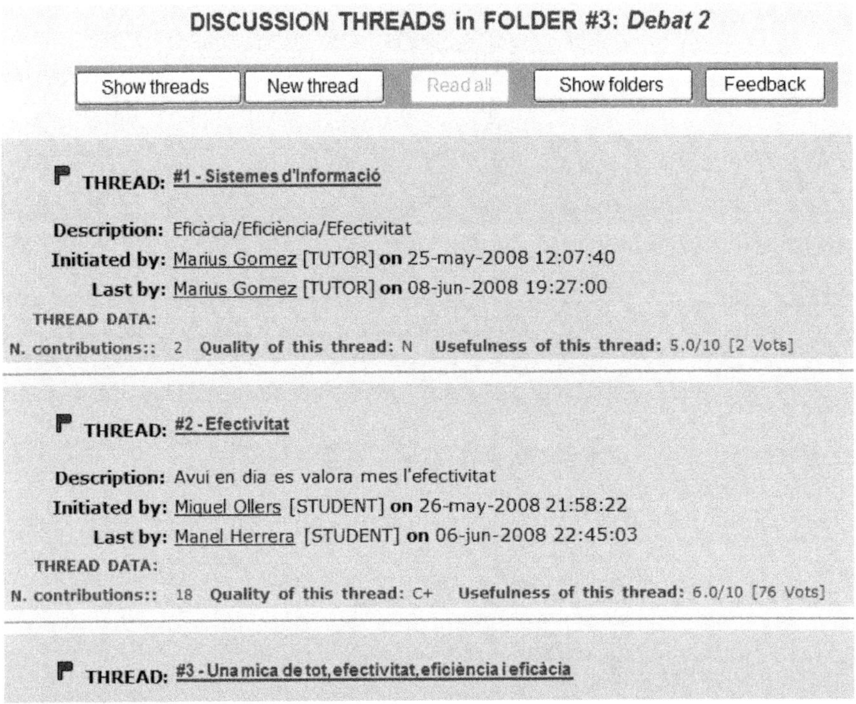

Fig. 4. A snapshot of some discussion threads inside a folder holding the discussion

CLPL's Functionality component provided suitable support in the design of the virtual places where the discussions take place. Indeed, the room entity was recursively used in different levels of abstractions, such as folders to hold the assignments featuring the class discussions and discussion threads inside each discussion. This also eased the implementation by reusing the same code for both purposes. This component also provided the suitable means to present the information of the knowledge acquired from the data interaction to the participants in the form of appropriate awareness and feedback, representing the cornerstone of this approach. The ultimate aim is to achieve a more effective interaction by allowing all participants to be aware of both their own and others' performance during the discussion process.

Therefore, the DF was especially designed to provide students with additional and important features to support the discussion in comparison to the traditional, well-known discussion tool used in the virtual classrooms of the UOC , such as (i) updated feedback, which includes the current mean number of all contributions' and complex indicators about the collaboration (see Figure 3) , (ii) threads in fully separated rooms (see Figure 4), (iii) open-closed branched dialogs (see Figures 5), and (iv) contribution qualifiers (see Figure 2).

Personal features of the discussion group participants (their role, collaboration preferences and so on) were taken into account and a user and group model were designed so as to allow participants to add new services whilst their needs evolve

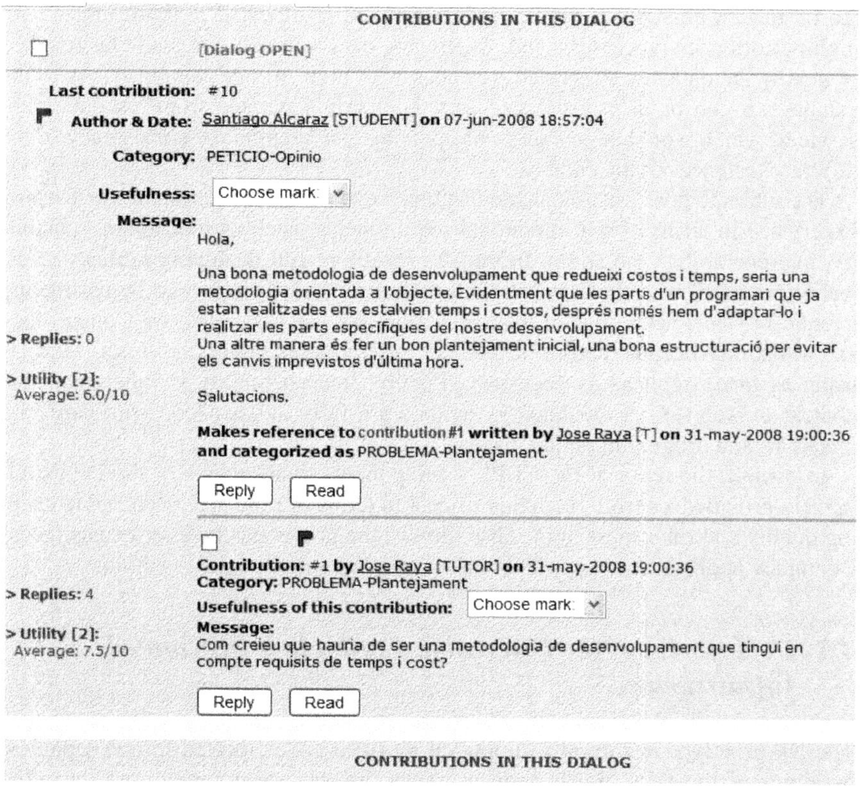

Fig. 5. Two dialogs in the same thread; upper dialog can be branched by checking the first checkbox

as the discussion moves forward. All these user features were included by the CSCL User Management component through the user profile management subsystem, providing a solid support for building and maintaining the user and group model.

Therefore, the DF constitutes a valuable learning resource that takes advantage of the CLPL to greatly improve essential features of a discussion process such as awareness of participant contributions. Indeed, by taking great advantage of the service-oriented approach of the CLPL functionalities, the primary principle during the DF implementation was to provide a broad set of independent fine-grained Web-services grouped by a particular purpose [5], such as the authentication process and the presentation of the feedback extracted.

A clear, independent, and separated vision of each single behavior of the DF into fine-grained task-specific web-services resulted in a natural distribution of the application into different nodes in a network.

This distribution was driven by matching the web-service purposes and the node configuration and location in the network. According to this view, the web services in the user interface layer should be allocated nearby the client;

the business web-services would be better suited if allocated in those nodes with high-performance processors, and, finally, the data web-services could be attached or nearby the database supported by nodes with high storage capability. As to the database, it can be also distributed as it is clearly separated from the data web-services, which would be in charge of updating and keeping the consistency of the different instances of the database.

The ultimate goal was to enhance and improve the effectiveness of the learning experience in terms of non-functional requirements, such as robustness, scalability, interoperability and so on. Indeed, by installing and deploying replicas of the web-services all over the network fault-tolerance is easily achieved by redirecting a request to an exact replica of the web-service when a node is down. Concurrency and scalability become natural in this context by parallelizing the users' requests using as many replicas as necessary. Finally, interoperability is inherent in the context of web-services technology as they are fully independent from hardware platforms and programming languages.

In overall, the reuse of the CLPL's components in all phases of the DF development provided a severe reduction of cost in terms of time and effort while keeping quality and robustness high. This allowed the provision and later extensions of a complex application to support collaborative learning in timely fashion.

4.3 Deployment of the Discussion Forum in a Distributed Infrastructure

The DF prototype is currently supported by three nodes located in two separated buildings of the UOC. Each node has very different configurations: Linux Red Hat 3.4.6-3 cluster, Intel Xeon CPU 3.00 GHz 4GB RAM; Windows 2003 server, Intel Pentium 3 CPU 800 MHz 512MB RAM; Linux SuSE 2.4.21-99 machine, Intel Pen-tium 4 CPU 2.00 GHz, 256MB RAM (see Figure 6).

For the purpose of our experience, all the Web-services of the DF prototype were replicated on each node. Moreover, the same client code in the form of PHP running on Apache Web servers was installed in two nodes (Windows server and Linux SuSE machine). Finally, in this prototype, just a single instance of the database was installed in the Windows server. This server acted also as an entry proxy by redirecting at HTTP level all the requests received to either itself or the Linux Red Hat cluster. In this first version the database is supported by just one node, which makes the system fully dependent from it. In future iterations it is planned to distribute the database in several nodes and manage its consistency by the data Web-services. The ultimate goal in this initial version was to prove the feasibility of the distributed approach.

To this end (see Figure 6), upon the reception of a user's request, the Windows server proxy first pings at Linux SuSE machine whether it is alive. If so, the Linux SuSE machine starts dealing with the request by executing its PHP code, otherwise the Windows server itself is doing so by executing the PHP code located in its own node. The client PHP code is actually in charge of starting the sequential call chain of Web-services for each layer, namely the user interface (UI), business

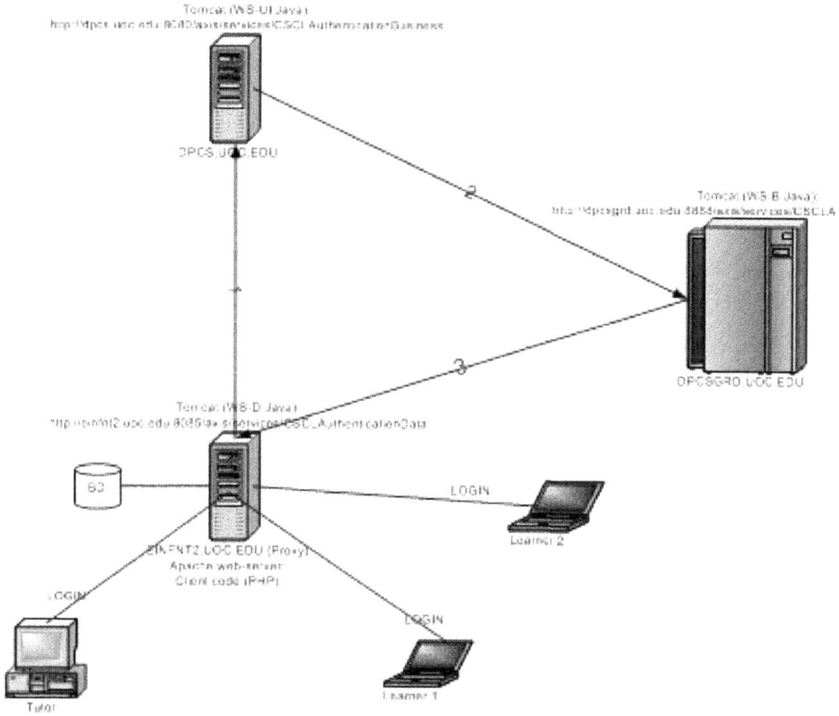

Fig. 6. Distribution of the DF functionalities into 3 nodes. Each node is in charge of a fine-grained Web-service representing a specific functional layer (i.e., user interface (UI), business (B), data (D)).

(B), and data (D) Web-services for each function requested. Thus, each Web-service call implies, if possible, to forward the current request to a different node. This means that before calling a Web-service on a different node a ping is always sent to check the node's availability. Whether the other two possible nodes are down, the node managing the current Web-service calls the next Web-service locally and tries again to find another node where to call the appropriate Web-service of the next layer. When the request finally arrives the data layer (i.e., the D Web-service), the call is addressed from any node to the Windows server. Once the information has been successfully managed in the database, the response is sent back to the client through the same request's way (i.e., same nodes and Web-services).

5 Computational Results and Evaluation

In order to validate the DF and analyze its benefits in the discussion process, two experiences were carried out at the UOC. Both experiences involved 40 graduated students enrolled in the course Methodology and Management of Computer

Science Projects in the last term. Each experience consisted of carrying out a discussion on a topic for 3 weeks involving all the students. The first experience was supported by using just one node (i.e., the Windows server) hosting the whole application, namely the Apache server managing the client's PHP code, all the Web-services and the database. In the second experience, our distributed approach was used.

In both cases, the discussion procedure was the same: each student was required to start a discussion thread by posting a contribution on the issue in hand, which resulted in as many threads as students. At the end of the discussion, each student was asked to close his/her thread with an improved contribution on the issue according to what s/he had learnt during the discussion. In the meantime, any student could contribute in both the own and any other discussion thread as many times as needed, as well as start extra threads to discuss new argumentations arisen. The aim was to evaluate the effect of the discussion process in the acquisition of knowledge of each student by comparing the quality of each thread's first and last contribution posted by the same student.

From the pedagogical point of view, the experience resulted very successful as it showed the benefits from providing an adequate information and knowledge management in supporting the discussion process. Indeed, the quantity and quality of the contributions during the discussion greatly increased in comparison to the experiences using the well-known but very poorly equipped asynchronous threaded discussion forum offered by the virtual campus of the UOC from the very beginning.

5.1 Results and Discussion

A statistical analysis of the results in the first experience comparing both the standard and the DF tools is shown in Table 1. Despite the standard tool generated more threads, most of them were actually empty (i.e. just 8 threads were contributed with more than 1 post vs. 42 threads in the DF). Moreover, the SD statistic for the posts/thread mean appears to be high in the DF, which proves the heterogeneity of the discussion involving threads of very different length. Note the very high SD statistic in the posts/students mean due to a single outlier, without which SD is 6,3. Finally, quality statistics are shown in terms of the number of words per contribution and the tutor assessment on the content. The higher number of number of words in the standards tool is due to the lack of discussion as most of threads were just started with a long opening contribution as a problem statement. On the other hand, the DF generated actual discussion and as a result the contributions became highly structured and specific. The tutor assessment row refers to content quality of all the contributions on average.

The qualitative evaluation of this first discussion was addressed by both examining those discussion threads that contained enough discussion (i.e. more than 7 posts) and checking whether the student who was in charge of each thread had posted both a start and close contribution on the same issue. The results on the DF showed that, in 28 threads fulfilling these requirements, 32% of students had improved their qualitative mark through the discussion in their threads, 68% kept the same mark, and no mark had dropped. On the other hand, no results were extracted from the discussion using the standard tool as it was poorly contributed; just 8 threads showed some discussion but only 4 had more than 7 posts.

However, during this first experience, many inconveniences arose due to the overuse of the Windows central server node by not only the participants of this experience but also many other students who carried out their learning activities, thus misusing this server as an academic resource. As a result, the discussion was interrupted several times due to the node's failures. Moreover, the discussion's participants suffered from serious lack of performance due to both the concurrency of different participants trying to gain access to the DF at the same time and the resource consumption of the server performed by external users. As a result, this generated a lot of frustration and complains about not being able to make progress on the discussion process.

Table 1. Main statistics extracted from the discussion using two discussion tools

Statistics	Standard tool	DF
Number of students	40	40
Number of threads	48	44
Total of posts	95	351
Mean number (posts/thread)	M=1,9 SD=2,4	M=7,9 SD=5,0
Mean number (posts/student)	M=2,3 SD=1,9	M=8,7 SD=8,1
Mean number (words/contribution)	M=352 SD=139	M=286 SD=85
Tutor assessment (average, out of 10)	7.2	7.6

Table 2. Excerpt of a questionnaire's results on the first experience using the DF tool supported by just one server

Selected questions	Average of structured responses (0 – 5)	Excerpt of students' comments
Asses the Discussion Forum (DF)	2	"Apart from serious technical problems, the DF fulfilled my expectations" "The system performed very slowly, I don't understand why the university is not able to provide us with a more powerful server!" "The DF is a powerful tool but most of times I couldn't even access because of timeout problems"
Evaluate how the DF fostered your active participation	3	
Did the DF help you acquire knowledge on the discussion's issue?	4	
Compare the DF to the campus' discussion standard tool	3	

Table 2 shows the results of a structured and qualitative report conducted at the end of the first experience addressed to the DF' users who were also asked to compare it to the standard well-known tool they had already used in previous discussions. This report shows the many technical problems occurring during the discussion due to the centralized approach.

The second experience was supported by the distributed version of the DF. Despite the functionality provided was the same as the previous experience, the results improved according to both the participants' and tutor' point of view. Indeed, the system performed smoothly and just one time the DF was reported to be unavailable. This improvement came mainly from the utilization of other nodes apart from the Windows server, which was still overused. This fact provided an important performance gain that all students appreciated a lot (see Table 3) and influenced on the discussion process in terms of participation impact and better quality in average (see Table 4).

Table 3 shows the results of the report conducted at the end of the second experience, which was the same as that conducted at the end of the previous experience.

Table 3. Excerpt of a questionnaire's results on the second experience using the distributed Discussion Forum tool

Selected questions	Average of structured responses (0 – 5)	Excerpt of students' comments
Asses the Discussion Forum (DF)	4	"The system performed much better and I could realize its potential"
Evaluate how the DF fostered your active participation	5	"Finally the technical problems seem to have been solved and I could participate at my pace"
Did the DF help you acquire knowledge on the discussion's issue?	5	"The statistical data and quality assessment displayed influenced my participation"
Compare the DF to the campus' discussion standard tool	4	"There is still more improvement to do as for the user interface but the system now performs well"

Table 4 shows a comparative study between the first and second experience. Certain key indicators, such as the tutor assessment and the participation impact, improved considerably, which show the benefits from the distribution approach in the learning process. Particularly interesting is the improvement of the passivity indicator showing the contributions on average pending to read. The reason may be found in the normalization of the system's performance, which allowed the participants to spend time reading others' contributions. This, in turn, enhanced the discussion process by increasing the cogniscitive level of the topic discussed.

Table 4. Main learning indicators extracted from both experiences

Indicators	First experience	Second experience
Tutor assessment 0-10 (on average)	6.2	7.8
Peer assessment 0-10 (on average)	5.4	6.5
Participation impact (on average)	+1.8	+4.1
Passivity (pending to read on average)	88.3%	31.9%

6 Conclusions and Future Work

This paper proposes a step further in the current software development methodologies by taking advantage of the most advance and latest techniques in software engineering, such as Generic Programming and Service-Oriented Architectures. The goal is to greatly improve software development in terms of quality, productivity and cost, as well to provide effective solutions to meet demanding non-functional requirements. To this end, an architectural solution in the form of a generic, service-oriented computational model called CLPL has been presented to help develop complex, modern and advanced e-Learning applications. Both the development experience of the CLPL and of a specific application, called Discussion Forum, based on this platform is reported to validate the key ideas proposed in this chapter.

The incorporation of distributed infrastructure on the DF has also been reported in order to meet certain important non-functional requirements that may influence the learning process a great deal, such as performance, scalability, fault-tolerance, and interoperability. The gain in performance might help, for instance, include more complex information of the collaboration to be generated and presented in real time (such as modeling the participants' behavior during the discussion by combining individual and group session and navigation information). Moreover, the benefits from having a flexible, interoperable environment makes it possible to completely integrate the DF system into the UOC's virtual campus. The aim of these improvements is both to enhance the effectiveness of complex collaborative learning processes (e.g., by avoiding a central point of failure) and stimulate the learning experience by describing and predicting students' actions and intentions as well as adapting the learning process to students' features, habits, interests, preferences, and so on.

We keep up exploring the benefits from using distributed infrastructure to support complex e-learning and to evolve appropriately the distributed version of the discussion tool prototype. However, this initial approach encourages us to work towards this direction. In the near future, it is planned to deal with the complex issue of distributing the database into the available nodes of the distributed infrastructure so as to avoid any central point of failure. Moreover, this initial approach is to be extrapolated by deploying the discussion tool in the nodes of PlanetLab platform so as to validate it in a real and complex distributed environment.

Acknowledgements

This work has been partially supported by the Spanish MCYT project TIN2008-01288/TSI. Fatos Xhafa's work is partially done at Birkbeck, University of London, on Leave from Technical University of Catalonia (Barcelona, Spain). His research is supported by a grant from the General Secretariat of Universities of the Ministry of Education, Spain.

References

[1] Amin, K., Nijsure, S., von Laszevski, G.: Open Collaborative Grid Services Architecture (OCGSA). In: Proceedings of the W3C EuroWeb 2002 Conference, Oxford, UK (2002)

[2] Bote Lorenzo, M.L., Gómez Sánchez, E., Vega Gorgojo, G., Dimitriadis, Y., Asensio Pérez, J.I., Jorrín Abellán, I.M.: Gridcole: a tailorable grid service based system that supports scripted collaborative learning. In: Computers and Education. Academic Press, Elsevier Ltd. (2007)

[3] Caballé, S., Xhafa, F.: Fostering Collaborative Knowledge Building by the Effective Provision of Knowledge about the Discussion Process. Special Issue on: Advances in Intelligent Information Management Systems and Applications. International Journal of Business Intelligence and Data Mining (IJBIDM) 4(2), 141–158 (2009)

[4] Caballé, S.: Combining Generic Programming and Service-Oriented Architectures for the Effective and Timely Development of Complex e-Learning Systems. In: Proceedings of the Second International Conference on Complex, Intelligent and Software Intensive Systems (CISIS 2008), Barcelona, Spain. IEEE Computer Society, Los Alamitos (2008)

[5] Caballé, S., Xhafa, F., Daradoumis, T.: A Service-oriented Platform for the Enhancement and Effectiveness of the Collaborative Learning Process in Distributed Environments. In: Meersman, R., Tari, Z. (eds.) OTM 2007, Part II. LNCS, vol. 4804, pp. 1280–1287. Springer, Heidelberg (2007)

[6] Caballé, S., Daradoumis, T., Xhafa, F.: A Generic Platform for the Systematic Construction of Knowledge-based Collaborative Learning Applications. In: Architecture Solutions for e-Learning Systems, ch. XII, pp. 219–242. Idea Group Inc. (IGI Global), Hershey (2007)

[7] Caballé, S., Paniagua, C., Xhafa, F., Daradoumis, T.: A Grid-aware implementation for Providing Effective Feedback to On-line Learning Groups. In: Meersman, R., Tari, Z., Herrero, P. (eds.) OTM-WS 2005. LNCS, vol. 3762, pp. 274–284. Springer, Heidelberg (2005)

[8] Caballé, S., Xhafa, F.: A Study into the Feasibility of Generic Programming for the Construction of Complex Software. In: Proceedings of the 5th GPCE/Net.Objectsdays 2003 (2003)

[9] Czarnecki, K., Eisenecker, U.W.: Generative Programming: Methods, Techniques, and Applications. Addison-Wesley, Reading (2000)

[10] Daradoumis, T., Martínez, A., Xhafa, F.: A Layered Framework for Evaluating Online Collaborative Learning Interactions. International Journal of Human-Computer Studies. Special Issue on Theoretical and Empirical Advances in Groupware Research 64(7), 1071–5819 (2006)

[11] Dillenbourg, P.: Introduction; What do you mean by "Collaborative Learning"? In: Dillenbourg, P. (ed.) Collaborative learning. Cognitive and computational approaches, pp. 1–19. Elsevier Science, Oxford (1999)
[12] Dimitriadis, Y.A., Asensio, J.I., Toquero, J., Estébanez, L., Martín, T.A., Martínez, A.: Towards a component based system for the CSCL domain (in Spanish). In: Proceedings of the Symposium on Informatics and Telecommunications, Sevilla, Spain (2002)
[13] Ellis, C., Gibbs, S., Rein, G.: Groupware: some issues and experiences. Communications of the ACM 34(1), 9–28 (1991)
[14] Foster, I., Kesselman, C., Tuecke, S.: The Anatomy of the Grid: Enabling Scalable Virtual Organizations. International Journal of Supercomputer Applications and High Performance Computing (2001)
[15] Gomaa, H.: Designing Software Product Lines with UML: From Use Cases to Pattern-Based Software Architectures. Addison-Wesley, Reading (2004)
[16] GuiLing, W., YuShun, L., ShengWen, Y., ChunYu, M., Xu, J., Meilin, S.: Service-Oriented Grid Architecture and Middleware Technologies for Collaborative E-Learning. In: IEEE SCC 2005, pp. 67–74 (2005)
[17] GuiLing, W., YuShun, L., ShengWen, Y., ChunYu, M., Xu, J., Meilin, S.: Towards a Grid-Based Collaborative Platform for E-Learning. In: Proceedings of WEBIST 2005, pp. 510–517 (2005)
[18] Gütl, C.: Moving Towards a Generic, Service-based Architecture for Flexible Teaching and Learning Activities. In: Pahl, C. (ed.) Architecture Solutions for E-Learning Systems, ch. I, pp. 1–24. Idea Group (IGI Global), Hershey (2007)
[19] Hew, K.F., Cheung, W.S.: Attracting student participation in asynchronous online discussions: A case study of peer facilitation. Academic Press, Elsevier Ltd. (2008) (in press)
[20] Inaba, A., Ikeda, M., Mizoguchi, R.: What Learning Patterns are Effective for a Learner's Growth? An ontological support for designing collaborative learning. In: Proceedings of the International Conference on Artificial Intelligence in Education, Sydney, Australia, pp. 219–226 (2003)
[21] Jonassen, D.H.: Learning with media: Restructuring the debate. Educational Technology Research and Development 42(2), 31–39 (1994)
[22] Koschmann, T.: Paradigm shifts and instructional technology. In: Koschmann, T. (ed.) CSCL: Theory and Practice of an Emerging Paradigm, pp. 1–23. Lawrence Erlbaum Associates, Mahwah (1996)
[23] Laforcade, P., Choquet, C., Nodenot, T., Pierre-Andre, C.: Model-Driven Engineering (MDE) and Model-Driven Architecture (MDA) applied to the Modeling and Deployment of Technology Enhanced Learning (TEL), Systems: promises, challenges and issues. In: Pahl, C. (ed.) Architecture Solutions for E-Learning Systems. IDEA Group Inc., Hershey (2007)
[24] OMG. Object Management Group: Model-Driven Architecture (2008), `http://www.omg.com/mda` (June 2009)
[25] Pankatrius, V., Vossen, G.: Towards E-Learning Grids: Using Grid Computing in Electronic Learning. In: Proceedings of IEEE Workshop on Knowledge Grid and Grid Intelligence, Halifax, New Scotia, Canada, pp. 4–15 (2003)
[26] Roseman, M., Greenberg, S.: Building Real Time Groupware with GroupKit, A Groupware Toolkit. ACM Transactions on Computer Human Interaction 3(1), 66–106 (1996)

[27] Simonson, M., Smaldino, S., Albright, M., Zvacek, S.: Teaching and Learning at a Distance. Prentice Hall, Upper Saddle River (2003)
[28] Soller, A.: Supporting Social Interaction in an Intelligent Collaborative Learning System. Int. J. of Artificial Intelligence in Education 12, 40–62 (2001)
[29] Stahl, G.: Group Cognition: Computer Support for Building Collaborative Knowledge. Acting with Technology Series. MIT Press, Cambridge (2006)
[30] Strijbos, J.-W., Martens, R., Prins, F., Jochems, W.: Content analysis: What are they talking about? Computers and Education 46(1), 29–48 (2006)
[31] W3C Working Group. Web Services Architecture Document (2004), `http://www.w3.org/TR/ws-arch/`
[32] Zumbach, J., Schönemann, J., Reimann, P.: Analyzing and Supporting Collaboration in Cooperative Computer-Mediated Communication. In: Koschmann, T., Suthers, D., Chan, T.W. (eds.) Computer Supported Collaborative Learning 2005: The Next 10 Years!, pp. 758–767. Lawrence Erlbaum, Mahwah (2005)

SISINE: A Negotiation Training Dedicated Multi-Player Role-Playing Platform Using Artificial Intelligence Skills

Kurosh Madani, Amine Chohra, Arash Bahrammirzaee, and Dalel Kanzari

Images, Signals and Intelligent Systems Laboratory (LISSI / EA 3956),
PARIS EST / PARIS 12 – Val de Marne University, Senart-FB Institute of Technology,
Bât. A – Av. Pierre point, 77127 Lieusaint, France

Abstract. In recent decades, a number of trainers have used role-playing games to teach negotiation skills. The SISINE Project – funded by the EU Leonardo Program - has developed a teaching methodology making it possible to conduct this kind of approaches in a virtual environment. The teaching methodology exploits a specially-developed technology platform allowing a small community of players to communicate, to interact and to play online in order to acquire basic notions and rules about negotiation and how to apply this knowledge. A part of SISINE project has investigated Artificial Intelligence issued techniques in order to evaluate implementation's possibility of computer-controlled "artificial players" embodying some intelligent behaviors. This chapter presents the first results of those investigations.

1 Introduction

There is a large literature showing that negotiation can reduce inter-individual conflict, and produce lasting solutions [1]. However, "*negotiation*" is a potentially exhausting and time-consuming complex process [2]. Encouraging groups to negotiate more frequently requires not only adequate education but also appropriated methodologies and tools. In recent decades, this need has motivated psychologists, educationalists and trainers to develop techniques for group problem solving, and methodologies to teach how to solve problems [3], [4]. One of the most popular among those developed techniques is *Psychodrama* [5] – a technique originally developed for psychotherapy. In this technique, each session is a sort of play. The therapist/director sketches out a story, assigns roles to the patients/actors, watches as they act out the story, and intervenes where this can be useful. At the end of the "play", the therapist leads a group discussion (e.g. debriefing).

When teachers use psychodrama for education, they talk about "role playing games". Here it is the teacher who takes on the role of the therapist and the learners who play the part of the patients. The literature describes the use of role-playing games in many different settings, from companies and trade unions to schools and political parties [6].

F. Xhafa et al. (Eds.): Computational Intelligence for Tech. Enhanced Learning, SCI 273, pp. 169–194.
springerlink.com

In recent years, the Internet explosion has led to the development of so-called ***M***assive ***M***ultiplayer ***O***n-line ***R***ole-Playing ***G***ames (MMORPG). Famous examples include Second Life (http://secondlife.com/education/) and Active Worlds http://www.activeworlds.com/edu/). "Massive" games like those aforementioned, allow large numbers of players to participate in a *virtual psychodrama*: remote users control a digital actor (e.g. an avatar) exploring a huge and hugely varied artificial universe, rendered in 3D graphics. Via their avatars, players take on specific roles, using the avatar as a mask in their interactions with other players. Players communicate using the possibilities provided by the avatars. Commonly, avatars can adopt a posture, make gestures or show specific facial expressions. They can also communicate, via text or speech [2]. In this way the remote user becomes a kind of "virtual" puppeteer. It has been suggested that players are less inhibited when playing electronic role playing games than they would be in real life. Divested of their social, psychological and corporeal identity, players find it easier to identify themselves in other lives (e.g. other roles) [7]. The effects can be both negative and positive. Excessive use can create disassociation and dependency; occasional, well-balanced use can stimulate players' skills allowing them to freely explore multiple scenarios [1]. The positive effects of video-games and their important role in the lives of young adults have encouraged researchers to investigate their use in education [8], [9].

This chapter deals with a platform allowing simple development and deployment of educational role-playing games aiming the negotiation skills' training. The platform has been called "SISINE" (*Sistema Integrato per la Simulazione e la Negoziazione).* The platform and the educational methodology on which it is based have been developed as part of the SISINE project, funded by the European Union "Leonardo" program"[1]. The SISINE software includes a set of specific possibilities for teachers allowing them to design training scenarios and a set of facilities for learners. The teacher's version includes tools allowing the teacher to define offline role-playing games for individual users; to define online multiplayer games; to intervene during multiplayer games; to manage debriefing sessions. The learner version allows learners to participate in single-player and multiplayer games and in debriefing sessions.

1.1 Motivations

Our motivation to link negotiation to e-Learning has been supplied by several points. One of them was that "*negotiation*" is a complex emotional interacting and decision-making process aiming to reach an agreement (compromise). Because

[1] National Research Council - Institute of Cognitive Sciences and Technologies (Italy) coordinator of the SISINE project. Laboratoire Images, Signaux & Systèmes Intelligents of Université PARIS 12 – Val de Marne (France), Entertainment Robotics (Denmark); ITTI - Institute of Communication and Information Technologies (Poland); Glaux, S.r.L. (Italy). TILS Spa (Italy); Mediazioni S.c.a.r.l. (Italy); PDCS – Partners for Democratic Chane Slovakia (Slovakia) and Xiwrite Srl (Italy). For further information see http://www.sisine.net.

"*negotiation*" is a potentially exhausting and time-consuming complex process, few people are effective negotiators. E-Learning and related technologies may be strong vectors in encouraging groups to train negotiation skills.

A second point was that if consequent efforts have been accomplished to remedy the leakage of negotiation training methodologies, the education of negotiation skills (and so, concerned educators) leaked modern tools to ease and reduce difficulty inherent to training of this complex and time-consuming process. E-Learning was a potentially privileged candidate to conceal this leakage, offering an additional appealing feature which is the "interactivity" (an intrinsic need as well in "*negotiation*" process as in its training methodologies).

Finally, the last critical point was related to diversity of domains concerned by the need of negotiation skills' training (as industrial, socio-economic, politic, medical, cultural, etc...). So, the design of a negotiation skills' training platform independent from the aforementioned diversity is a real advantage. Again, an E-Learning based approach appeared suitable to handle this diversity.

1.2 Related Work

Concerning this chapter, SISINE project and the presented work, there is not really what could be directly called "related work". However, at the same time and looking from different slat, a large spectrum of works may relate the chapter's subject. In fact, if a large number of works has concerned the "negotiation's dynamics modeling", "E-Learning and related innovative technologies" and "Agent or Multi-Agent" topics, works combining "negotiation's practice" training, "E-learning" and MMORPG using innovative information technologies remain inexistent.

Over the past decade a number of negotiation dedicate systems have been proposed (see [10], [11], [12], [13] and [14]). If the accomplished works have demonstrated some significant improvements, each of them has concerned a specific negotiation domain. Moreover, most of presented works has been relevant to architectural aspects and leaded to moderate implementations (e.g. software dedicated to concepts' validation aiming essentially an internal usage). On the other hand, either the user's preferences have been assumed to be known, or preference elicitation techniques have been incorporated in order to cover a same very specific need. An example is the Althena project (see www.althenasoft.org). In fact, in the issued implementation, the user has to build all content models by himself (or herself). This simulation oriented support does neither include predefined structures nor library of models nor interaction support. Moreover, the training dedicated skills are totally omitted.

Probably, the most relative work to a part of the contents of the present chapter is the recent "Pocket Negotiator" project, proposed in 2008 and financed by Netherlands Organization for Scientific Research (www.onderzoekinformatie.nl /en/oi/nod/onderzoek/OND1334832/). The project aims to design and implement a human-machine collaborative system in which user and machine explicitly share a generic task model. The applicative goal is to support human negotiator in coping with emotions and moods in human-human interactions. However, again the project deals with a somehow specific aspect of human-human negotiation and seems to focus integrative bargaining. Anyway, here also, the training and e-learning dimension is out of the purpose.

1.3 Chapter's Concerns Andorganization

Part of SISINE project has investigated possibilities' evaluation of "intelligent" artificial negotiator's implementation (e.g. computer-controlled avatars imitating a number of intelligent behaviors). The present chapter deals with this part of the SISINE project, describing the investigated directions and related results. A first goal was to provide computer-controlled avatars some unpredictable (by opposition to deterministic rules based predictable) computer-controlled personality (e.g. character or behavior). However, our investigations have trailed further purposes: as those related to potentially autonomous computer-controlled "player". If the aforementioned ambitious aim opened a large number of possible investigation directions, our contribution mainly concerned the Artificial Intelligence (AI) based approaches and especially those related to "bio-inspired" AI: machine learning, soft-computing, etc...

The rest of this chapter is organized as follows: the next section will introduce the SISINE platform, its environment and the offered possibilities in order to understand the SISNE avatar's (e.g. artificial negotiator's) general structure. The section 3 will describe the simple deterministic model used in first beta version of SISINE platform to implement the computer-controlled avatar. In the section 4, we will introduce possible architectures of an un-deterministic avatar (e.g. by opposition to "predictable"), based on Q-Learning paradigm. This section will also enlighten simulation results and will compare the behavior of a Q-Learning based avatar and the deterministic one. In the section 5 will introduce a first level of intelligent agent using machine learning skills (e.g. Q-Learning based and ANN based ones). This section will also enlighten simulation results and will compare the behavior of a machine learning avatar (e.g. Q-Learning based and ANN based ones) and the deterministic one. Finally the last section will conclude the chapter and will open future directions of our investigations.

2 SISINE and Its Environment

Inspired by "Role Playing Games", SISINE is a virtual environment offering a set of especially dedicated skills for negotiation process simulation offering a Multi-player On-line Role-Playing Games (MORPG) platform for negotiation training (teaching). In fact, a chief trait distinguishes SISINE from other role playing games: beside the functions normally available in any role playing game, SISINE provides "distance learning" oriented additional specially designed facilities. This chief specificity makes a SISINE online training session less free than a traditional chat session: learners are less free to interact than they would be in a traditional chat session. In fact, the script (e.g. negotiation frame, objectives and scenario) is written by the teacher. Players have to take on specific roles and pursue specific objectives described in the script.

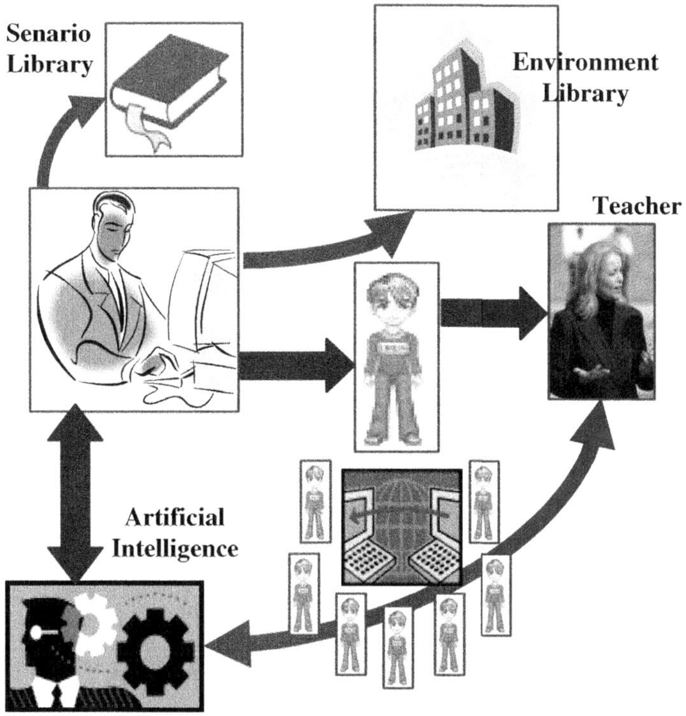

Fig. 1. Bloc diagram of SISINE architecture

SISINE provides a complete set of tools for teachers allowing them either to write scripts for online multiplayer games or for single player exercises (so-called "gyms"). When designing a multiplayer game, the teacher can choose the roles, the goals and the bodies of individual players. Once the game is in progress a teacher can observe what is going on, intervening at any moment, writing messages to players or activating special "events". And when it is over the teacher can lead a group discussion analyzing, criticizing or correcting the adopted strategies.

To define scripts and the role of specific actors (players), teacher use the SISINE "Game Editor". The editor allows, among other possibilities, to define constraints in order to give a direction to the game play. Fig. 1 gives the bloc diagram of SISINE platform's architecture showing main bloc-functions of its architecture. Script writing takes place in five phases:

1. General description of the negotiation story.
2. Definition of the "actors" (name, description, personal objectives, avatar)
3. Definition of "private actor characteristics" (i.e. characteristics visible only to the learner controlling the actor).
4. Definition of document events to be incorporated in the script
5. Definition of message events to be incorporated in the script

Fig.2 and Fig.3 show examples of different screens with their various windows and menus available within the editor to define the script's title, the general

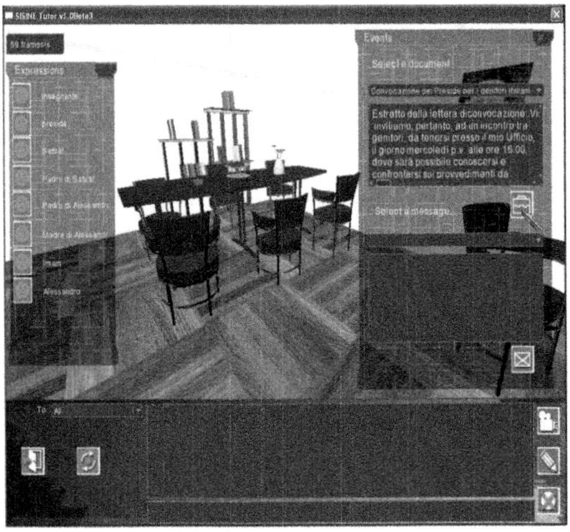

Fig. 2. Example of a script-writing: allowing to define the graphics environment, the background to the "negotiation scenario" and the players' (e.g. negotiators') shared "objectives"

description of the story, the graphic environment associated to the story, and the players' "shared objectives".

Once the script has been created, the teacher associates each "actor" (negotiator) with a specific learner. As soon as learners log on, they take on the role of the actor, chosen for them by the teacher. As well as preparing the scripts for online games and assigning actors to users, there are two other ways in which teachers can intervene in learners' interactions with SISINE. One is to take on the role of one of the actors. The other is to act as an invisible stage director. In this second role, teacher can:

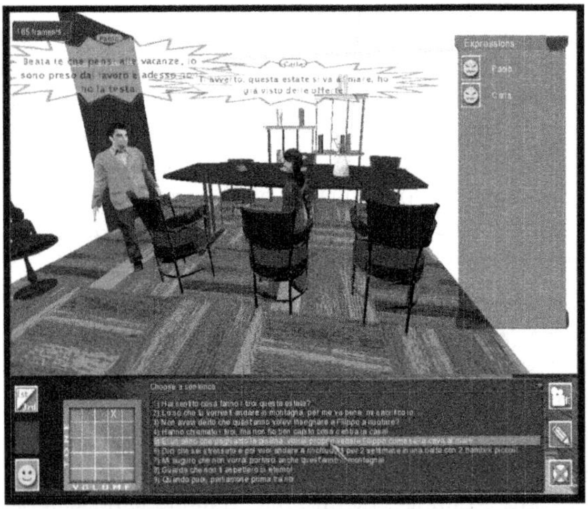

Fig. 3. Screen hard copy showing a negotiation game's sequence

Once the script has been created, the teacher associates each "actor" (negotiator) with a specific learner. As soon as learners log on, they take on the role of the actor, chosen for them by the teacher. As well as preparing the scripts for online games and assigning actors to users, there are two other ways in which teachers can intervene in learners' interactions with SISINE. One is to take on the role of one of the actors in the simulation. The other is to act as an invisible stage director. In this second role, teacher can:

- invisibly observe the interactions among the players
- access the actors' "private characteristics"
- listen in to private messages ("whispers") between players
- "broadcast" messages visible to all players
- exchange private messages with a specific user
- activate events, changing the course of the simulation

To do all this, teacher uses the tutor module. SISINE allows teacher to record both multiplayer games and gyms. The recordings take the form of a 3D interactive video. SISINE provides possibilities to interact with recordings in several different ways. In particular one can:

- view the recording from different directions, shifting the virtual camera to the slant they prefer,
- view the script for the session,
- insert comments,
- change the speed of the recording
- split the recording into chapters (as on a DVD)
- modify the recording in various ways (e.g. cutting out less interesting parts)

Through these facilities, teachers can create a commented version of the recording for use during *debriefing*.

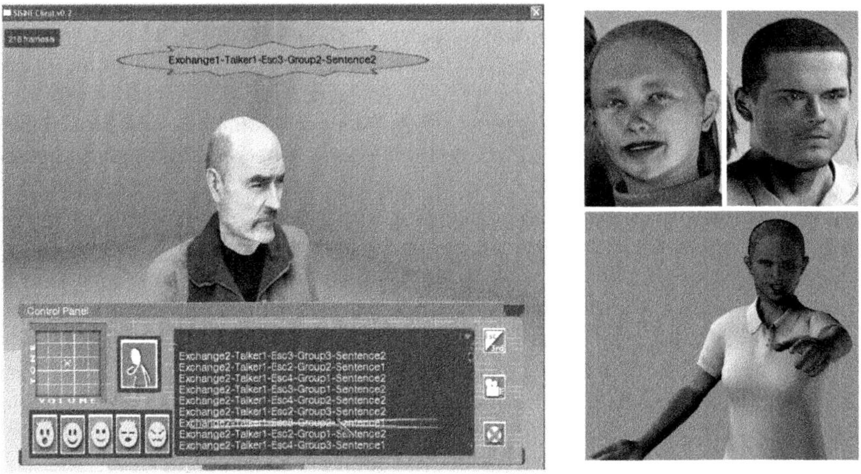

Fig. 4. (left) Screen hard copy showing an avatar and the control panel during a game sequence. (right) Examples of avatars with their facial and gesture expressions.

Fig. 5. Screen hard copy showing a negotiation game sequence in 2nd SISINE beta-version

As well as interacting with other players in online multiplayer games, SISINE learners can interact with computer-controlled actors in offline games and exercises (so-called "gyms"). The gyms allow learners to practice strategies they can later apply to their interactions with human players. As in multiplayer games, it is the teacher who prepares the scripts for exercises. However, the scripts for the gyms are more detailed, specifying not only the basic story and the actors involved (one controlled by the learner the other by the computer) but the set of phrases the actors can use to communicate with each other, the "para-verbal" signals (volume, tone of voice, facial expressions, gestures) associated with each phase and the effect of the phrase on the other actor (e.g. increased or decreased belligerence). To control the artificial actors (e.g. the computer-controlled negotiator's character), SISINE uses techniques either based on deterministic rules or issued from AI.

The current beta version supports up to twenty simultaneous players, represented by avatars. Communication among players is based on short text messages displayed in bubble cartoons above the avatars' heads. Special controls allow players to control the avatars' movements, gestures and facial expressions. In this way, the community (players) can communicate, interact and negotiate (play) online. Fig. 4 shows example of an avatar during a negotiation game sequence as well as the related control window. The right side of the same figure gives examples of avatars and their gestures and facial expressions. Since January 2007, the SISINE project has been testing its platform and methodology in Poland, Slovakia and Italy. In Poland the experimental group consisted of company sales representatives; in Slovakia of managers in NGOs; In Italy of teachers (e.g. social problems between teachers, children and the children's parents). In each case, the groups have used a custom-designed virtual environment to practice specific forms of negotiation: commercial negotiation (in Poland), negotiation in human resources management (Slovakia) and intercultural negotiation (Italy). Fig. 5 shows the screen hardcopy of the 2nd SISINE beta-version corresponding to a negotiation game's sequence involving two players (teacher and child in an Italian school).

3 Negotiation Approach Based on Escalation Level and Negotiator's Personality

3.1 Model of the Negotiation Process between Two Negotiator Agents

The suggested negotiation approach relies on a theoretical model of the negotiation process (between two negotiator agents) which is mainly based first on the escalation level (defining gradually several negotiation stages from agreement to interruption) of the negotiation. Second, it is based on the negotiator personality, i.e., characters Conciliatory (Con), Neutral (Neu), and Aggressive (Agg) which define a "psychological" aspect of the negotiator agent personality.

In this theoretical model, seven possible escalation level stages are modeled by the variable escalation level EscLevel belonging to the interval [0, 60] and numbered from 0 to 6 as follows: (0). agreement if [0, 10 [; (1). defense of positions [10, 19], where each part defends a position and attempts to persuade the other of its validity ; (2). intermediate level stage [20, 29] ; (3). attack on the other's position [30, 39], where each part do not discuss its position but only seek to attack the other's position ; (4). intermediate level stage [40, 49] ; (5). attack on the other [50, 59], where each part do not discuss the problem but attack each other ; (6). interruption of the negotiation] 59, 60], ending the negotiation process.

Also, in this theoretical model, the character of a negotiating agent (negotiator) is defined by a character vector [Con, Neu, Agg] where each component belongs to the interval [0, 100] in percentage (%) such as the character vector components verify Eq. (1):

$$Con + Neu + Agg = 100(\%) \tag{1}$$

A "bot" (avatar) could have three characters:

- "Aggressive" (Agg) corresponding to an "irritable" bot who might "lose its temper" very quickly,
- "Neutral" (Neu) corresponding to a calmer behaviour making such a bot to be more willing to look for an agreement,
- "Conciliating" (Con) corresponding to a peaceful bot who might accept the point of view of the other player.

Taking into account the aforementioned, the bot's character could be represented as a 3-D (3-componets) vector where components represent the part (percentage) of the coolness (e.g. conciliating posture), permissiveness (neutrality) and

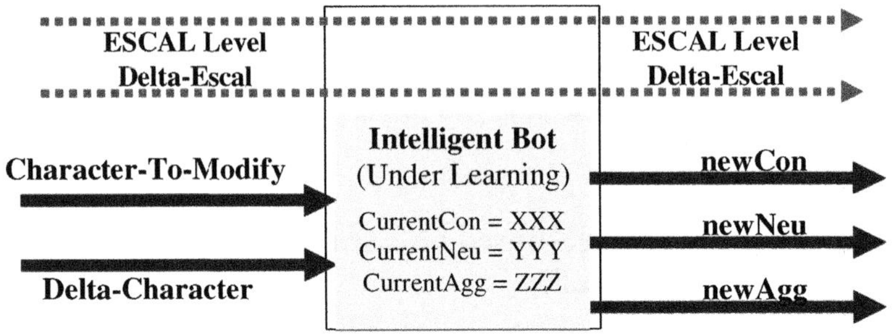

Fig. 6 Black-Box model of computer-controlled bot's character

nervousness (aggressive ingredient) of its character, respectively. Consequently, during a negotiation round, each negotiator agent is defined by its current character vector **CurrentChar** = [CurrentCon, CurrentNeu, CurrentAgg], then, its new character vector **NewChar** = [NewCon, NewNeu, NewAgg] is updated from the user sentence effect explained below.

So, the model generating the SISINE computer-controlled bot's character could be considered as a black-box updating the bot's character (e.g. updating the values of the three aforementioned character-vector's components). The character-vector is updated taking into account the actual bot's character and the challenger's action. The challenger's action (inputs of the Black-box) are represented by two parameters: "Character-To-Modify" (an "alphanumerical parameter which value belongs to the ensemble {"con" , "neu" , "agg"}) and "Delta-Character" which value belongs to the interval [-100 , +100]. The choice of such interval has essentially been guided by "computational" convenience. The first one depends to the sense of the phrase pronounced by the challenger (stimulating one of the three above-indicated parameters) and the second to the solemnity of the pronounced phrase (its excitatory or inhibitory influence on the stimulated parameter). In other words, the bot's temperament (and so, its reaction) will evolve as a function of its actual character and the challenger's action. Fig. 6 shows the bloc-diagram of such a Black-Box.

3.2 Deterministic Model of SISINE Avatar's Character

The deterministic updating rules fashioning the artificial bot's character are quite simple. Depending on the two previously-mentioned inputs, they are based on weighted modification of the three character-vector's components (character's parameters). The updated parameters (newCon, newNeu and newAgg) are computed according to a general expression ruling the computer-controlled deterministic bot's new character expressed by the Eq. (2), where:

new-Char $\in$ {newCon, newNeu , newAgg}

Current-Char$\in$ {Current-Con, Current-Neu, Current-Agg}

Delta-Char = (Delta-Character / 100) **X** Current-Char
Delta-Character $\in$ [-100 , +100]
$\alpha_j \in \Re$ and $j \in \{1,2,3\}$

$$\textbf{new-Char} = \textbf{Current-Char} + (\alpha_j \times \textbf{Delta-Char}) \tag{2}$$

The above-given set of artificial bot's character updating rules could be interpreted as boosting the bot's new character according to the "Character-To-Modify": the character-vector's component corresponding to "Character-To-Modify" is reinforced and the two other components are impaired. However, depending on nature of the "Character-To-Modify" those two different components are weakened differently. In fact, if bot's coolness is boosted (e.g. Character-To-Modify = "con") then the priority is given to decrease the bot's nervousness. Wile, if the bot's permissiveness is reinforced (e.g. Character-To-Modify = "neu") then the two other traits are diluted equally. Finally, if the aggressive personality is privileged (e.g. Character-To-Modify = "agg") then the precedence tends to shrink the conciliating posture.

4 Q-Learning Based SISINE Avatar Character's Model

As a first level of "intelligence" we settled on awarding the artificial bot some learning ability. Prior to expect more sophisticated (higher level intelligent) behavior, the primal goal was to make the artificial bot's character some how less predictable than in the case of the deterministic updating rules based computer-controlled version. However, the objective of reaching a more sophisticated level of intelligence (by integrating the learning ability) has not been disregarded.

The training ability has based on the *reinforcement learning* paradigm, and more especially, on *Q-Learning* variant of this approach. Our choice has been motivated on the one hand by relative simplicity of such learning process and on the other hand by its "goal-directed" nature: learning what to do (how to map situations to actions) so as to maximize a reward signal. In fact, in this kind of artificial learning process, instead to learn a number of predefined actions to take (stated by an expert as being the best actions to take) as it is done in most forms of supervised-learning based machine learning, the *reinforcement learning* based system has to discover which actions yield the most reward by trying them.

In the most interesting and challenging cases, actions may affect not only the immediate reward but also the next situation and, through that, all subsequent rewards. These two characteristics, trial-and-error search and delayed reward, are the two most important distinguishing features of reinforcement learning [15], [16]. Concerning "how" the computer-controlled bot could acquire "intelligent behavior" from learning ability, it will be discussed soon after: within the two subsections devoted to the suggested *Q-Learning* based bot character's controlling strategies.

4.1 Q-Learning: Principle and Advantages

One of the most widely used variants of reinforcement learning is *Q-Learning* [15], [16], [17]. In this on-line reinforcement learning paradigm, the agent incrementally learns an action/value function Q(s, a) that it uses to evaluate the helpfulness of performing action *a* while in state *s*. *Q-Learning* leads to optimal behavior, i.e., behavior that maximizes the overall utility (performance) for the agent in a particular task environment [18]. Indeed, the policy used to generate behavior, called the *behavior policy*, may in fact be unrelated to the policy that is evaluated and improved, called the *estimation policy*. In other words, from the interaction with an environment an agent can learn, using Q-Learning, to maximize the reward leading to an optimal behavior policy. The Fig. 7 illustrates this agent-environment interaction from which an agent can learn, using Q-Learning, to maximize the reward leading to an optimal behavior policy.

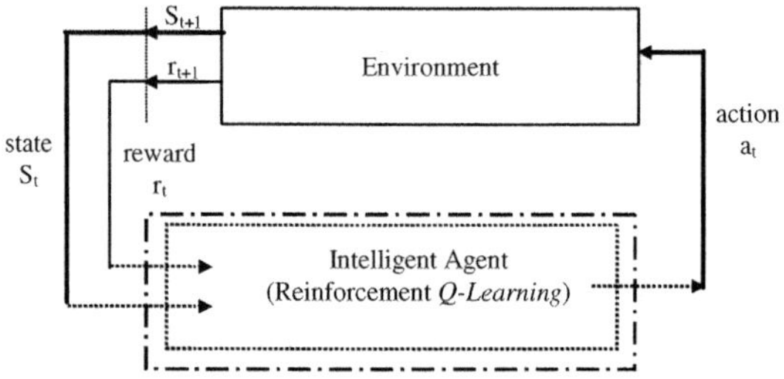

Fig. 7 Bloc-diagram on a *Q-Learning* based agent interacting with an environment

Reinforcement learning approaches offer two important advantages over classical dynamic programming [19]. First, the approaches are on-line, i.e., they have the capability to take into account the dynamics nature of real environments. This permits them to focus their attention on the parts of the state space that are important and to ignore the rest of the space. Second, the approaches can employ function approximation techniques, e.g., NN, to represent their knowledge. This allows them to generalize across the state space so that the learning time scales much better.

4.2 First Level Q-Learning Based Un-deterministic Bot

As a first tentative of some bot's intelligent behavior, we have introduced the learning ability using *Q-Learning* model. The goal at this first level of intelligence is to make the bot's character more unpredictable than in the case of the deterministic one: we will denote the bot including such ability by "First-Level Intelligent Bot" (FLIB). In fact, the ability to learn (more or less) a given "character" leads to a more un-deterministic bot revealing a more un-predictable character. The input parameters are the same that those of the deterministic bot (Character-To-Modify and Delta-Character). As the deterministic bot, the *Q-Learning* based bot updates its character (e.g. computes $[newCon , newNeu , newAgg]^T$) using the reinforcement learning process. Fig. 8 gives the bloc-diagram of a FLIB interacting with a human (human-controlled avatar) or with a computer-controlled avatar such bot. Receiving its input parameters from a human challenger (or anther bot) and starting from a special initial character (none-character, represented by Initial CurrentCon = 33.33, Initial CurrentNeu = 33.33 and Initial CurrentAgg = 33.33), the learning module generates the new character-vector ($[newCon , newNeu , newAgg]^T$) awarding the reinforcement of the required traits and punishing the reinforcement of the two other components.

Example 1:
Purpose: to make bot acquire a conciliatory character (e.g. a conciliatory bot).

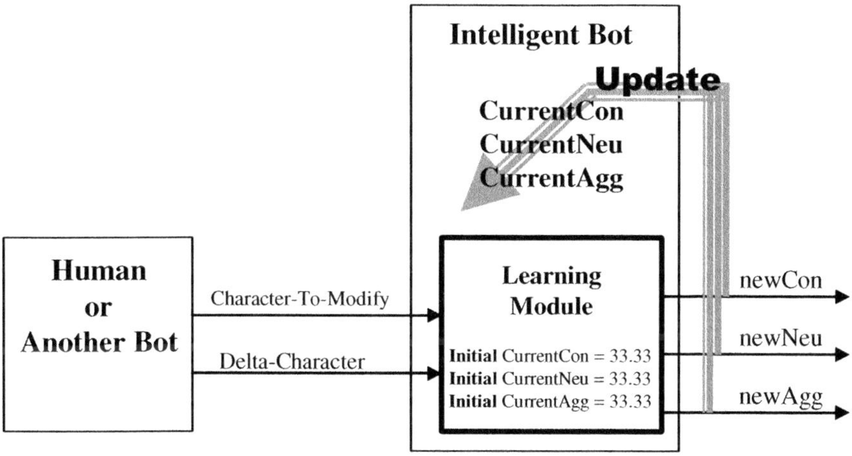

Fig. 8. Bloc-diagram of the proposed Q-*Learning* based personality updating mechanism

Input parameters: Character-To-Modify = "con"
Delta-Character = 20

Results: newCon = 38.72
newNeu = 30.55
newAgg = 30.55

Comments & interpretation: Starting from a "weak character" (e.g. CurrentCon = 33.33, CurrentNeu = 33.33, CurrentAgg = 33.33), the bot acquired a Conciliatory character.

Another applicative variant of this strategy is a human controlled avatar with adjustable character. In this case, contrary to the last version where the number of learning iterations is stipulated (by bots' designer), the number of learning loops (iterations) is controlled by the human player (negotiator). In such way, the negotiator represented by such an avatar (incorporating the FLIB) will be able to fashion (to control) the avatars character, preserving some "unpredictability" of the avatar's behavior. In other words, the negotiator (human) will repeat a single-step

Q-Learning based process as many times as it will be necessary until reaching the expected personality.

4.3 Second Level Q-Learning Based Un-deterministic Bot

Another implemented variant of this strategy is to update the bot's current character according to only the most dominant traits of the identified character. Fig. 9 gives the bloc diagram corresponding to a SLIB using this variant of character imitation (adaptation) ability. The main difference between the two bots is the current character-vector's updating rule. In the previous case, the current character-vector is replaced by the new character-vector. While here, the bot's current character-vector is updated according to the *Q-Learning* based procedure given by Fig. 9.

This variant of character imitation leads to a bot adapting its character to the most dominant trait of another challenger involved in the negotiation sequence. This means that the character identification is performed only to identify the most dominant component and estimate the corresponding Delta-Character parameter. Then a *Q-Learning* based procedure (similar to the architecture shown in Fig. 8)) updates the current character-vector. Probably, the most significant advantage using the above-described character adaptation strategy is related to the fact that the resulted (adapted) character is globally similar to the dominant traits of the

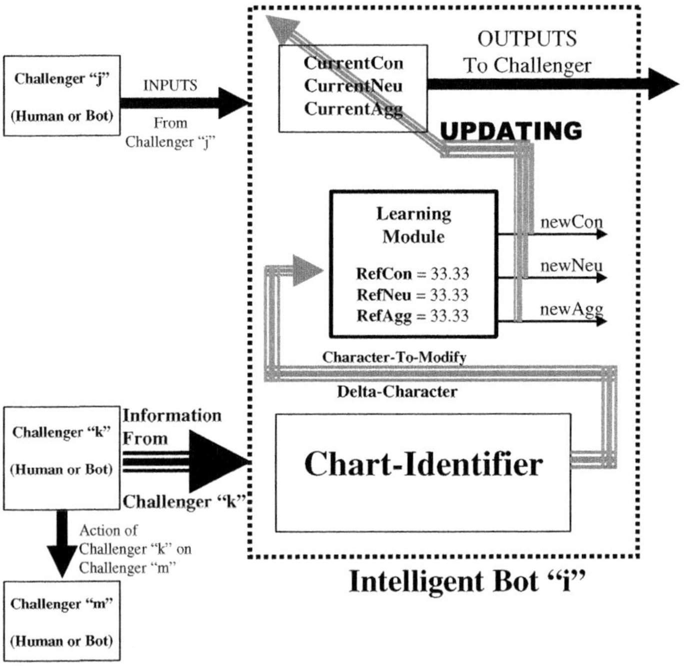

Fig. 9. Bloc-diagram of the SLIB operating according to the second character adaptation variant

identified character without looking-like it exactly. The next case study (example) compares the updated character using this variant with the previous variant.

Example 2:
Purpose: taking for granted that the "k" bot (human or computer-controlled player) is characterized by the charter-vector $[Con , Neu , Agg]^T = [35.00 , 40.00, 25.00]^T$, identify the character of the bot "k" and generate new character-vector in order to adapt (update) the bot's current character according to the identified traits using the strategy depicted by Fig. 9.

RefCon = 33.33
RefNeu = 33.33
RefAgg = 33.33

Input parameters: TargetCon = 35.00
TargetNeu = 40.00
TargetAgg = 25.00

Results:
Step-1: identification of the most dominant chart
Results
Character-To-Modify = "neu"
Delta-Character = 20.01
Step-2: learning the predominant character
Results
Character-To-Modify = "neu"
Delta-Character = 20.01
newCon = 30.55
newNeu = 38.73
newAgg = 30.55

Comments & interpretation: After adaptation, the bot's character becomes as well neutral as the imitated character but remains globally different from that (imitated) character.

Probably another potential advantage of this variant is inherent to the fact that the reference character here could be also the bot's current character leading to different unpredictable character of the bot. Let us consider the next example, where the reference character is different from the previously considered one (e.g. RefCon $\neq$ 33.33, RefNeu $\neq$ 33.33, RefAgg $\neq$ 33.33).

Example 3:
Purpose: assuming that the "i" bot's current character is $[Con , Neu , Agg]^T = [25.00 , 30.00, 45.00]^T$ and taking for granted that the "k" bot (human or computer-controlled player) is characterized by the charter-vector $[Con , Neu , Agg]^T = [35.00 , 40.00, 25.00]^T$, identify the character of the bot "k" and generate new character-vector in order to adapt (update) the "i" bot's current character.

RefCon = 25.00
RefNeu = 30.00
RefAgg = 45.00

Input parameters: TargetCon = 35.00
TargetNeu = 40.00
TargetAgg = 25.00

Results:
Step-1: identification of the most dominant chart
Results
Character-To-Modify = "neu"
Delta-Character = 20.01
Step-2: learning the predominant character
Results
Character-To-Modify = "neu"
Delta-Character = 20.01
newCon = 22.50
newNeu = 34.85
newAgg = 42.50

Comments & interpretation: After adaptation, the bot's character becomes globally less aggressive and more neutral but totally different from the imitated character.

It is pertinent to note that a large set of rules could be designed (implemented) leading to more or less different resulted adapted (imitated) characters. Our aim here is not to present and analyze all possible strategies but to show and interpret the most appropriated of them for our purpose. One of these additional slants is the fusion strategy. It consists in fusing information related to identified (imitated) character and features related to current bot's character. We have implemented a variant of such approach fusing two identified Delta-Character parameters: one related to the target character and the other related to the bot's own current character. The fusion policy is quite simple: the Delta-Character parameter used to update the bot's character is obtained by adding the two identified Delta-Character parameters. The idea here is to correct (compensate) the target Delta-Character parameter by the Delta-Character required to acquire the bot's current character.

Example 4:
Purpose: in the frame of the same conditions as those of the previous example (Example 3), adapt the bot "i" character using the above-described fusion based imitation strategy.

RefCon = 25.00
RefNeu = 30.00
RefAgg = 45.00

Input parameters: TargetCon = 35.00
TargetNeu = 40.00
TargetAgg = 25.00

Results:
Step-1: identification of the most dominant
target character

Results

Character-To-Modify = "neu"

Delta-Character = 20.01

Step-2: identification of the most dominant current character

Results

Character-To-Modify = "agg"

Delta-Character = 35.01

Step-2: learning the predominant character

Results

Character-To-Modify = "neu"

Delta-Character = 20.01 + 35.01 =55.02

newCon = 18.14

newNeu = 43.35

newAgg = 38.14

Comments & interpretation: After adaptation, the bot's character becomes globally neutral and close to the imitated dominant traits. However, concerning the aggressive traits the bot's aggressive component's new value (newAgg = 38.14) is far from the imitated character's one (TargetAgg = 25.00) remaining relatively close to its initial aggressive traits. The interpretation of this case could be done as follow: contrary to the imitated bot (who reveals a peaceful neutral state), the current bot remains sensible (aggressive) even if it adapts a globally neutral behavior.

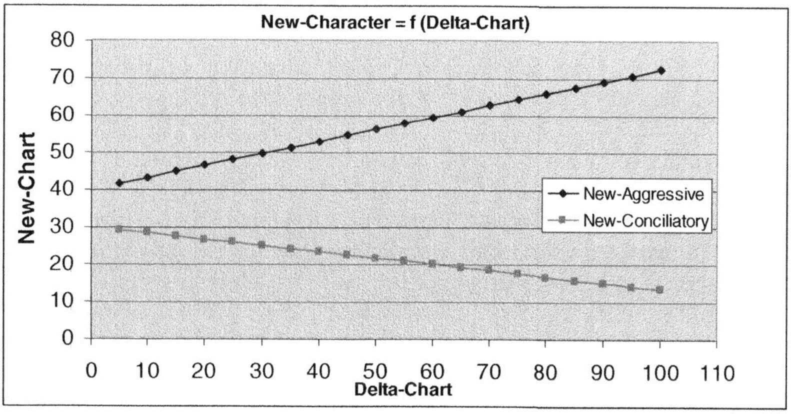

Fig. 9. Q-Learning based updated character versus Delta-Character parameter

If the introduction of reinforcement based learning mechanism, either in its so-called "first level" (FLIB) or "second level" (SLIB) of intelligence, leads more-or-less to the expected behavior, the evolution of the issued bot's character remains some-how linear (especially in the case of the FLIB). Moreover, in actually used *Q-Learning* based personality updating policy (Fig. 8), if the privileged character

(character to be modified) becomes somehow erratic, the two other components remain affected equally. Fig. 10 gives bot's updated character's evolution (e.g. new character-vector's components: newCon , newNeu and newAgg) versus Delta-Character parameter, when the used updating is performed conformably to the policy illustrated by Fig.8. This figure has been obtained assuming the following scenario: starting from the current character $[30\ ,\ 30\ ,\ 40]^T$ (e.g. CurrentCon = 30, CurrentNeu = 30 and CurrentAgg = 40), the computer-controlled bot acquires progressively aggressive personality (Delta-Character varying from 5 to 100).

5 First Level of Intelligence: Q-Learning and Neural Network Based Agents

As it has been mentioned and used above, one of the most important breakthroughs in reinforcement learning was the development of an off-policy temporal-difference control algorithm known as Q-learning [20], [21], [22]. Elsewhere, Association of ANN module with Q-Learning offers additional advantages of the generalization quality and limited memory requirement for storing the knowledge [26], [27]. Intrinsically, ANNs are characterized by their learning, and generalization capabilities (firsts essential traits of intelligent behaviors' emergence), robustness, massively parallel computational potentiality and distributed memory skills [22], [23], [24]. Here, two agents are compared versus their characters' (e.g. personalities') evolutions based on the learning action: one based on Q-Learning (as in Fig. 8) and the other based on association of ANN module with Q-Learning. Figure 11 gives the intelligent agent's personality acquiring module's bloc diagram. The non linearity inherent to the ANN's learning process, as for example of an ANN with Back-Propagation learning mechanism (e.g. MLP-like ANN [25], [26]) may confer the bot a learning based "qualitative" change of its character. However, the success of this kind of associative (and supervised) learning processes is conditioned by availability of a reliable (realistic) training database covering the expected behavior.

In fact, values of the Delta-Character may symbolize here some range of "qualitative" character's variation requirements. Applicative use of this in SISINE platform consists to adjust (increase or decrease) the traits by using a natural language (e.g. "more aggressive", "less neutral", etc...) instead to perform the adjustment by numerical values of the parameters. The use of such qualitative information makes the bot's behavior's control as well as it's interaction with "human negotiator" more realistic. However, the Delta-Character's range has been normalized as belonging to [-10 , +10] interval (instead of [-100 , +100]).

Delta-Character = -1: "decrease weak"
Delta-Character = +1: "increase weak"
Delta-Character = -4: "decrease middle"
Delta-Character = +4: "increase strong"
Delta-Character = -6: "decrease strong"
Delta-Character = +6: "increase strong"

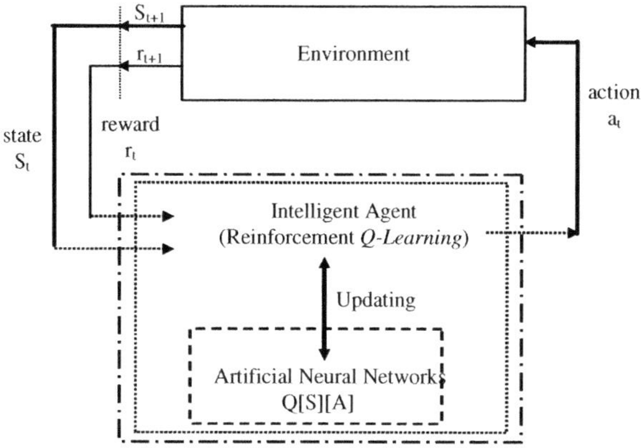

Fig. 11. Bloc-diagram on a *Q-Learning* based agent interacting with an environment. (right) Bloc-diagram on a *Q-Learning* based intelligent agent incorporating an ANN interacting with an environment.

(a)
```
Initialize Q(s, a) arbitrarily
Repeat (for each episode):
  Initialize s (initial state)
  Repeat (for each step of episode):
      Choose a from s using policy derived from Q
      Take action a, observe r, s'
      Q(s, a) ← Q(s, a) + α[r + γmax_a' Q(s', a') − Q(s, a)]
      s ← s'
  until s is terminal (final state)
```

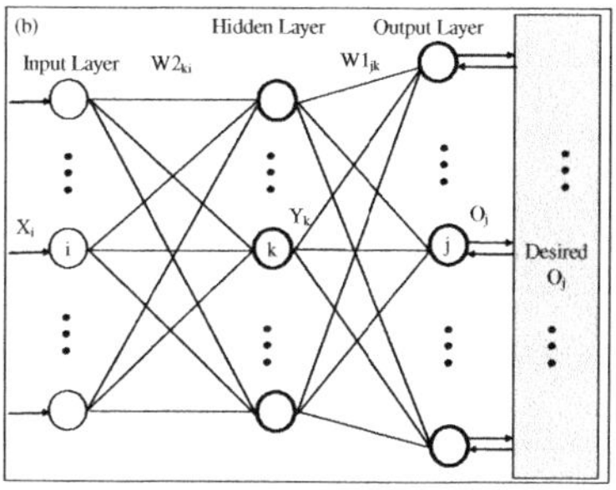

Fig. 12. Q-Learning: an off-policy temporal-difference control learning paradigm (a) and ANN's architecture where X_i (i = 1, ..., 5), Y_k (k = 1, ..., 11), and O_j (j = 1, ..., 3), (b)

The NN architecture is built of three layers input layer, hidden layer, and output layer as shown in Fig. 12 (b). The input vector is **X** = [CurrentCon, CurrentNeu, CurrentAgg, CharacterToModify, DeltaChar]. These components are then transformed according to the set of relations given in Eq. (3), to constitute input vector **X**.

$$
\begin{aligned}
X_1 &= (1/\rho)\exp(-\text{CurrentCon}/a),\\
X_2 &= (1/\rho)\exp(-\text{CurrentNeu}/a),\\
X_3 &= (1/\rho)\exp(-\text{CurrentAggt}/a),\\
X_4 &= (1/\rho)\exp(-\text{CharToModify}/a),\\
X_5 &= (1/\rho)\exp(-\text{DeltaChar}/a),
\end{aligned} \tag{3}
$$

where ρ: norm of input vector **X** and a: input pre-processing factor with a > 1.

Input Layer: This layer is input layer with i input nodes receiving the components of the input vector **X**. This layer transmits inputs to all nodes of next layer.

Hidden Layer: This layer is the hidden layer with k hidden nodes. The output of each node is obtained using the output sigmoïd function f as follows:

$$\mathrm{net}_k = \sum_i X_i \, W2_{ki} \text{ , and } Y_k = f(\mathrm{net}_k)\text{, where } f(x) = \frac{1}{1+\exp(-x)}. \tag{4}$$

Output Layer: This layer is the output layer with j linear output nodes obtained by:

$$O_j = \sum_k Y_k W1_{jk} \tag{5}$$

The steps in the used learning algorithm are outlined as follows:

1- Random weight ($W2_{ki}$ and $W1_{jk}$) initialization [-1, +1].
2- Apply an input vector **X** to the input layer.
3- Compute net_k and outputs Y_k of the hidden layer.
4- Compute outputs O_j of the output layer.
5- Compute the error δ_j for the outputs of the output layer:

$$\delta_j = (\mathrm{Desired}O_j - O_j). \tag{6}$$

6- Compute the error δ_k for the outputs of the hidden layer:

$$\delta_k = f'(\mathrm{net}_k) \sum_j \delta_j W1_{jk} \text{ , and } \delta_k = Y_k (1 - Y_k) \sum_j \delta_j W1_{jk} \text{ ,}$$
$$\text{since f: sigmoïd function} \Rightarrow f' = f\,(1 - f). \tag{7}$$

7- Update the weights of the output layer:

$$W1_{jk}(t+1) = W1_{jk}(t) + \Delta W1_{jk}, \text{ with } \Delta W1_{jk} = \eta\ \delta_j\ Y_k. \tag{8}$$

8- Update the weights of the hidden layer:

$$W2_{ki}(t+1) = W2_{ki}(t) + \Delta W2_{ki}, \text{ with } \Delta W2_{ki} = \eta\ \delta_k\ X_i \tag{9}$$

9- Compute the error E:

$$E = (1/2) \sum_j (\mathrm{Desired}O_j - O_j)^2 \tag{10}$$

10- Repeat 2- to 9- with the same input vector **X** (the same training example) until the error E is very close to the tolerance.
11- Repeat 2- to 10- for each input vector **X** (each training example).
12- Repeat 2- to 11- under several epochs.

The used Q-learning paradigm is shown in Fig. 12 (a). The parameter settings of the initial Q values, the constant step-size parameter ($0 < \alpha <= 1$), and the discount rate ($0 < \gamma <= 1$) have been done following the choice approaches given in [11] and [12] resulting in: initial Q values = 0.5, $\alpha = 0.1$, and $\gamma = 0.01$.

For both agents (e.g. the Q-Learning only and the Q-Learning/ANN association based one) the input vector is **X** = [CurrentCon, CurrentNeu, CurrentAgg, CharacterToModify, DeltaChar]. The goal is to generate the output vector **O** = [NewCon, NewNeu, NewAgg], corresponding to the "character change" for different values of "DeltaChar" (e.g. DeltaChar taking values -6, -4, -1, +1, +4 and +6, respectively). Using the Fig.8 diagram's based architecture, two sets including each 144 situations have been generated. One of the two sets using to train the ANN and the other for its test.

Diagrams of Fig. 13 give the training and testing results and comparing each of them to the Q-Learning based case. The left diagram corresponding to the learning phase has been obtained using the starting input vector **X** = [CurrentCon = 38.00, CurrentNeu = 31.50, CurrentAgg = 29.50, CharacterToModify = Con, DeltaChar]. It is pertinent to note that "CharacterToModify = Con" means that the character's component to be modified is "conciliatory" part of the global character's state. While, for the right diagram corresponding to the testing phase, a different

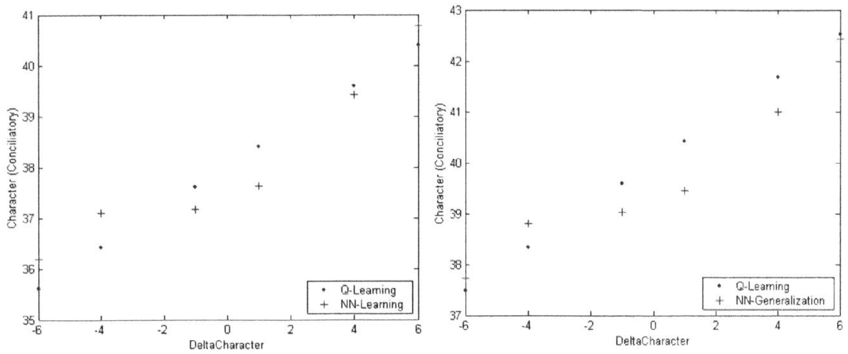

Fig. 13. Example of NewCon component evolution in ANN-learning phase (left) and ANN-testing phase (right)

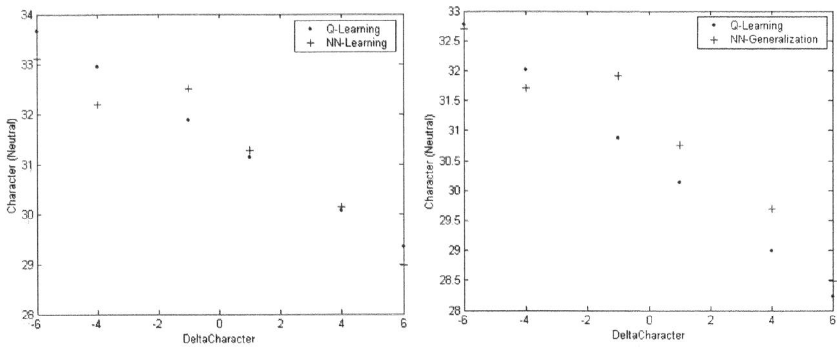

Fig. 14. Evolution of NewNeu component within the Fig.12 example in ANN-learning phase (left) and in ANN testing phase (right)

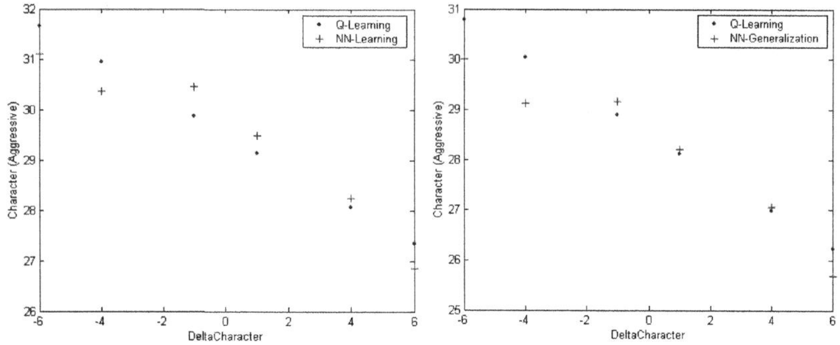

Fig. 15. Evolution of NewAgg component within the Fig.12 example in ANN-learning phase (left) and in ANN testing phase (right)

unlearned initial input vector has been considered : **X** = [CurrentCon = 40.00, CurrentNeu = 30.50, CurrentAgg = 28.50, CharacterToModify = Con, DeltaChar]. Of course, as for the previous case, here the character's component to be modified is also "conciliatory" part of the global character's state. Figures 14 and 15 give the evolution of the two other character's components (e.g. NewNeu and NewAgg) within the same example.

6 Discussion and Conclusion

Inspired by "Role Playing Games", SISINE provides a complete set of tools for teachers allowing them either to write the scripts for online multiplayer games or for single player exercises (so-called "gyms"). In this chapter, we have presented and discussed a number of possible strategies and issued schemes, based on artificial learning mechanisms in order to make the computer-controlled player's character more unpredictable (more realistic). We proposed and analyzed a number of slants (essentially based on *Q-Learning* and ANN paradigms) in order to realize (implement) a more-or-less unpredictable character for the SISINE computer-controlled bot. If the introduction of reinforcement based learning mechanism, either in its so-called "first level" (FLIB) or "second level" (SLIB), leads more-or-less to the expected behavior, the evolution of the issued bot's character remains some-how linear (especially in the case of the FLIB). The use of a non linear learning process, as for example Artificial Neural Networks (ANN) with Back-Propagation learning mechanism confers the bot a non linear change of its character. Simulation results (Figures 12, 13 and 14) confirmed this fact.

The toy-simulation shows also that one can operate using qualitative information making the bot's behavior's control more realistic: associating a semantic representation to "traits' magnitudes" uprising (or decreasing), for example by using a lookup table. As a first level of "intelligence" we settled on awarding the artificial bot some learning ability. Prior to expect more sophisticated (higher level intelligent) behavior, the primal goal was to make the artificial bot's character some how less predictable than in the case of the deterministic updating rules based computer-controlled version. However, the objective of reaching a more sophisticated level of intelligence (by integrating the learning ability) has not been disregarded.

Another point meriting completing this discussion is the application testing of the SISINE platform. A part of project has concerned the SISINE trial based on educational experiment. Three countries (partners) have been involved in this phase: Italy, Slovakia and Poland. Different negotiation scenarios (scripts) have been designed and integrated in SISINE platform. The Italian educational experiment, conducted by TILS, has concerned intercultural negotiation and focused the frame of a children-teacher conflict in Italian schools (elementary, middle and high schools). It has been realized involving 16 participates (one Head-master of school and 15 teachers originated from the three above-indicated levels). The Slovak pilot-study, conducted by PDCS, concerned social negotiation aiming group decision-making. 6 negotiators (participants) have taken part in this experimantation. The polish trial, conducted by ITTI, concerned group decision-making as well within a buisiness negotiation frame and has involved 7 participants. The main remarks (results) issued from the afforementioned trials could be summerized in following points:

- Regarding the software (which have been used by all participants) and its different possibilities (functions) :
  - 100% of participants used group simulation;
  - 100% of participants used the "relational training" with an artificial partner (e.g. computer controlled deterministic bot);
  - Only 20% of participants used training of an artificial, "intelligent" partner (bot);
  - Only 10% of participants used session recording and subsequent visualization skills;
  -
- Regarding the participants' feeling in the use of platform's commands have been:
  - 90% found message sending easy;
  - 90% found control of the avatar movements quite complicated;
  - 80% found the commands related to group session quite easy to use;
  - 80% found the commands of relational training and "bot" training understandable and quite easy to use.
- Regarding the already implemented scripts and negotiations' skills training:
  - 80% of the participants stated that objectives, stories and roles of the characters were described clearly. The remaining 20% said that they had difficulties in understanding the objectives.
  - 75% of the participants felt that the object of negotiation was interesting, whereas the remaining 25% found that the object of negotiation was not very convincing and therefore it was not possible to conduct a satisfying negotiation.

- In general the majority of participants in the experimentation felt that stories and objectives of the characters were clearly described and interesting. The characters were able to convey valid lessons on negotiation.
- In Poland the participants where very supportive to the idea of teaching negotiation techniques via an e-learning platform. According to the participants such courses allow the development of negotiation skills which are necessary in everyday life. Despite the shortcomings pointed out previously, all participants unanimously expressed their wish of participating in similar courses again.
- The tutor's role has been acknowledged as fundamental by 100% of the participants. The presence of tutors has obviously facilitated understanding the game and facilitated the negotiation process through reflections and comments.

Concerning the impact of number of users (negotiators) on SISINE platform, the aforementioned experimentations have not been appropriated to study this aspect. However, based on hardware related considerations, it seems natural that the hardware's characteristics (the used computers, the network's speed and band pass, etc…) will have a key impact on number of users and the quality of training (fast or slow interaction with tutor as well as with negotiation group).

Actually we are working on analysis and implementation of more complex ANN based strategies in order to increase the bot behavior's non linearity. At the same time, we evolve the agent's (e.g. computer controlled negotiator's) behavior by embedding additional functions. One of these additional functions is a fuzzy inference based decision module, which will allow the artificial negotiator to choose the most appropriated response (action) against its challenger. However, in order to control the integration of such additional functions, we have restricted the negotiation scenarios (e.g. the negotiation environment) to the "seller-buyer" negotiation case.

Acknowledgments. Authors would like to express their gratitude to ***the EU Leonardo Program*** for supporting SISINE project and the above-reported investigations. We sincerely thank our Italian colleagues from CNR di Roma, as well as other partners involved in SISINE project, for their collaboration. We also acknowledge Dr; Veronique Amarger, Dr. Abdennasser Chebira, and Dr. Christophe Sabourin from our laboratory for fruitful discussions.

References

[1] Miglino, O., Di Ferdinando, A., Rega, A., Benincasa, B.: SISINE: Teaching negotiation through a Multiplayer Online Role Playing Game. In: Proceedings of 6th European Conference on e-Learning (ECEL 2007), Copenhagen, Denmark, pp. 439–448 (2007)

[2] Fox, W.M.: Effective Group Problem Solving: How to Broaden Participation, Improve Decision Making, and Increase Commitment to Action. Jossey-Bass Inc. Pub. (1987)

[3] Qin, Z., Johnson, D.W., Johnson, R.T.: Cooperative versus Competitive Efforts and Problem Solving. Review of Educational Research 65(2), 129–143 (1995)

[4] Lewicki, R.J.: Teaching Negotiation and Dispute Resolution in Colleges of Business: The State of the Practice. Negotiation Journal 13(3), 253–269 (1997)
[5] Moreno, J.L., Zelany, L.D.: Role theory and sociodrama. In: Roucek, J.S. (ed.) Contemporary Sociology. Philosophical Library, New York (1958)
[6] Asherman, I.G., Asherman, S.: 25 Role Plays to Teach Negotiation, vol. 2. HRD Press (2003)
[7] Fleming Seay, A., Kraut, R.E.: Project massive: self-regulation and problematic use of online gaming. In: Proceedings of the SIGCHI conference on Human factors in computing systems, pp. 829–838. ACM Press, New York (2007)
[8] De Freits, S., Griffths, M.: Online gaming as an educational tool in learning and training. British Journal of Educational Technology 38(3), 535–537 (2007)
[9] Miglino, O., Di Ferdinando, A., Rega, A., Ponticorvo, M.: Le nuove macchine per apprendere: simulazioni al computer, robot e videogiochi multi-utente. Alcuni prototipi *Sistemi* Intelligenti 18(1), 103–136 (2007)
[10] Gutman, R., Maes, P.: Agent-mediated Integrative Negotiation for Retail Electronic Commerce. In: Noriega, P., Sierra, C. (eds.) AMET 1998 and AMEC 1998. LNCS (LNAI), vol. 1571, pp. 70–90. Springer, Heidelberg (1999)
[11] Jonker, C.M., Treur, J.: An Agent Architecture for Multi-Attribute Negotiation. In: Proceedings of the 17th International Joint Conference on AI, IJCAI 2001, pp. 1195–1201. Morgan Kaufman, San Francisco (2001)
[12] Klein, M., Faratin, P., Sayama, H., Bar-Yam, Y.: Protocols for Negotiating Complex Contracts. IEEE Intelligent Systems Journal, special issue on Agents and Markets 18(6), 32–38 (2003)
[13] Luo, X., Jennings, N.R., Shadbolt, N.: Acquiring user strategies and preferences for negotiating agents: a default then adjust method. International Journal of Human Computer Studies 64(4), 304–321 (2006)
[14] Jonker, C.M., Robu, V.: Automated Multi-Attribute Negotiation with Efficient Use of Incomplete Preference Information. In: Proceedings of the Third International Joint Conference on Autonomous Agents and Multi-Agent Systems (AAMAS 2004), pp. 1056–1063. IEEE Computer Society Press, Los Alamitos (2005)
[15] Whitehead, S.D.: Reinforcement Learning for the Adaptive Control of Perception and Action. Technical Report 406, Department of Computer Science, University of Rochester (1992)
[16] Sutton, R.S., Barto, A.G.: Reinforcement Learning. MIT Press, Cambridge (1998)
[17] Watkins, C.J.C.H.: Learning from Delayed Rewards. Ph.D Thesis, King's College (1989)
[18] Pfeifer, R., Scheier, C.: Understanding Intelligence. MIT Press, Cambridge (1999)
[19] Dietterich, T.G.: Hierarchical reinforcement learning with the MAXQ value function decomposition. Journal of Artificial Intelligence Research 13, 227–303 (2000)
[20] Rumelhart, D., Hinton, G., Williams, R.: Learning Internal Representations by Error Propagation. In: Rumelhart, D., McClelland, J. (eds.) Parallel Distributed Processing: Explorations in the Microstructure of Cognition, I & II. MIT Press, Cambridge (1986)
[21] Arbib, M.A. (ed.): Handbook of Brain Theory and Neural Networks, 2nd edn. MIT Press, Cambridge (2003)
[22] Anderson, J.A.: An Introduction to Neural Networks. The MIT Press, England (1995)
[23] Patterson, D.W.: Artificial Neural Networks: Theory and Applications. Prentice-Hall, Simon & Schuster (Asia) Pte Ltd., Singapore (1996)

[24] Haykin, S.: Neural Networks: A Comprehensive Foundation, 2nd edn. Prentice-Hall, Englewood Cliffs (1999)
[25] Sandholm, T.W.: Distributed Rational Decision Making. In: Multiagent Systems: A Modern Introduction to Distributed Artificial Intelligence, pp. 201–258. MIT Press, Cambridge (1999)
[26] Lin, L.-J.: Self-improving reactive agents based on reinforcement learning, planning and teaching. In: Machine Learning, vol. 8, pp. 293–321. Kluwer Academic Publishers, Dordrecht (1992)
[27] Touzet, C.F.: Neural reinforcement learning for behaviour synthesis. Robotics and Autonomous Systems 22, 251–281 (1997)

Computational Intelligence Methods for Data Analysis and Mining of eLearning Activities

Pavla Dráždilová, Gamila Obadi, Kateřina Slaninová, Shawki Al-Dubaee, Jan Martinovič, and Václav Snášel

FEECS, VŠB - Technical University of Ostrava,
Department of Computer Science,
Ostrava, Czech Republic
pavla.drazdilova@vsb.cz, gobadi@hotmail.com,
slaninova@opf.slu.cz, shawkialdubaee@gmail.com,
jan.martinovic@vsb.cz, vaclav.snasel@vsb.cz

Abstract. Enhancing the the effectiveness of web-based eduction has become one of the most important concerns within both educational engineering and information system fields. The development of information technologies has contributed to the growth in elearning as an important education method. This learning environment enables learners to participate in 'any time, any place' personalized training. It has been known that the application of data mining and computational intelligent approaches can provide better learning environments, and in their effort to participate in this field, the authors introduced this study which consists in its first part of a survey of the applications of data mining and computational intelligence in web based education during (2004–2009), and the second part is a case study that aims to analyze students' activities performed in a Learning Management System.

1 Introduction

Enhancing the the effectiveness of web-based eduction has become one of the most important concerns within both educational engineering and information system fields. The development of information technologies has contributed to the growth in elearning as an important education method. This learning environment enables learners to participate in 'any time, any place' personalized training. It has been known that the application of data mining and computational intelligent approaches can provide better learning environments. Fuzzy logic based methods are used for evaluation of students [44, 82] and other techniques such as association rules or multiagent systems were used for classification and personalization [34, 53]. Due to the large amount of information generated by elearning systems researches have been interested with data clustering [22] . The authors are interested in data mining and computational intelligence methods related to processing, analysing and evaluating students activities records generated from a learning management system [60]. In this work they evaluated students activities, identified irregular models of study behaviour and applied some visualization techniques on the students data. In this

F. Xhafa et al. (Eds.): Computational Intelligence for Tech. Enhanced Learning, SCI 273, pp. 195–224.
springerlink.com

chapter, the authors pursue their interest in the field of data mining and computational intelligence and its applications in elearning. The first part of the chapter is a survey of different applications of data mining and computational intelligence in web based education during (2004–2009), and the second part is a case study that aims to analyze students' activities performed in a Learning Management System.

2 Learning Managment Systems

eLearning is a method of education that utilizes a wide spectrum of technologies, mainly internet or computer-based, in the learning process. (LMS) - also called Course management systems (CMS) or Virtual learning environment (VLE) systems - provide effective maintenance of particular courses and facilitate communication between educators and students and within the student community. These systems usually support the distribution of study materials to students; content building of courses, preparation of quizzes and assignments, discussions, distance management of classes. In addition, these systems provide a number of collaborative learning tools such as forums, chats, news, file storage etc.

Unlike conventional face-to-face education methods, computer and web-based education environments provide storage of large amounts of accessible information. These systems record all the information about students' actions and interactions onto log files or databases. Within these records, data about students learning habits can be found including favored reading materials, note taking styles, tests and quizzes, ways of carrying out various tasks, communicating with other students in virtual classes using chat, forum, and etc. Other common data, such as personal information about students and educators (user profiles), student results and user interaction data, is also available in the systems' databases.

This collected data is essential for analyzing students' behavior and can be very useful in providing feedback both to students and educators. For students, this can be achieved through various recommended systems and through course adaptation based on student learning behavior. For teachers, some benefits would include the ability to evaluate the courses and the learning materials, as well as to detect the typical learning behaviors.

Regardless of its benefits, the huge amount of information generated by learning management systems makes it too difficult to manage these data collections and to extract useful information from them. To overcome this problem some LMS offer basic reporting tools, but in such large amounts of information the outputs become quite obscure and unclear. In addition, they do not provide specific information of student activities while evaluating the structure and contents of the course and its effectiveness for the learning process [119]. The most effective solution to this problem is to use data mining techniques.

3 Data Mining Techniques in Web-Based Educational Systems

There are various definitions of data mining presented in research area. In [35] is presented that Data mining is the nontrivial extraction of implicit, previously unknown, and potentially useful information from data. Benoit [15] offers the definition of relative discipline Knowledge Discovery in Databases (KDD), which he refers to as data mining: Data mining is a multistage process of extracting previously unanticipated knowledge from large databases, and applying the results to decision making. Data mining tools detect patterns from the data and infer associations and rules from them. The extracted information may then be applied to prediction or classification models by identifying relations within the data records or between databases. Those patterns and rules can then guide decision making and forecast the effects of those decisions. In Principle of Data Mining [41] is presented: Data mining is the analysis of (often large) observational data sets to find unsuspected relationships and to summarize the data in novel ways that are both understandable and useful to the data owner.

Many researchers presented basic phases of data mining. Detailed description of particular phases is stated in [32]: Learning the application domain, Creating a target dataset, Data cleaning and preprocessing, Data reduction and projection, Choosing the function of data mining, Choosing the data mining algorithm(s), Data mining, Interpretation and Using discovered knowledge. More pregnant description of data mining phases is mentioned by Schuman [78]: Collection of data, Data preprocessing, Data analysis, Data visualization and Data interpretation. In the same paper is proposed unified definition of Data Mining and novel interesting view to Data mining phases with relation to chemical states of aggregation.

Another field of data mining research is web searching and information retrieval from the World Wide Web. The research is oriented to data mining of logs available from search engines for further recommendation of results or queries. Query recommendation of related queries for search engine users using past queries stored in large-scale web access logs and web page archive is presented in [56]. Method for weighted social networks construction based on information on the web and search engines such as Google is described in [55]. Some search engines have integrated social bookmarking services, individual search history information, or statistics of search activities (e.g. Google [103], Google Trends [104], Yahoo Buzz, Ask IQ [105]).

3.1 Data Mining and eLearning

Data mining is a multidisciplinary area. It uses several methods to build analytic models that discover interesting patterns and tendencies from obtained data collections. The eLearning data mining process consists of the same phases as found in general data mining processes [75]:

- Data collection - LMS produce a large amount of information about student activities and interaction during the learning process. These data collections are stored using databases or log files.
- Data preprocessing - Data collections are cleaned and transformed into an appropriate format to be used in the data mining application phase.

- Data mining application - Usage of data mining techniques and algorithms to obtain the required information, data summaries of discovered knowledge and visualizations of mined information or data models in relation to the requirements of users (instructors, students, system administrators).
- Data interpretation and result implementation - Data mining results are interpreted and used by educators or students to improve student learning processes (LMS).

The application of data mining in eLearning is an iterative cycle [73]. This thought is based on the fact that creating an eLearning entails a complicated and demanding process. The course developer (teacher or instructor) must design a course structure and its components in a way that ensures suitability of a given course's character while fulfilling student study requirements and providing different means for communication during lessons. Based on data obtained from LMS, student activity during the term may be monitored. At the end of a term, study results and course effectivity may be evaluated and any necessary improvements may be made. Data mining results are often applied for adapting courses to user profiles and study assessments. Another area of applying data mining techniques involves the collaborative learning process, in which students create a community for sharing information about criteria for completing courses successfully [26].

The classification of data mining techniques and computational intelligence methods is well known and has been mentioned in several publications [23, 2].

Each e-learning system stores a large amount of data based on the history of the users interactions with the system. Such dataset is a good source of very useful knowledge. With the aid of data mining techniques we can analyze data and create patterns and data groups. Such obtained models can be successfully applied to make the learning process more effective by adding more functionalities to the system such as: personalization of the learning process, Feedback for authors of educational content, building students groups and detect the degree of plagiarism in students homework.

Data mining for e-learning is more complicated than the other applications of of data mining (for example e-business). Students in an elearning course may access a lesson many times, browse other sites and other parts of the course, leave and return, play games in between, and work over long periods of time. Defining and capturing students' activities from the system logs is a major challenge. Furthermore, linking this data with existing data about learners (students' profiles) can give better information, but requires integrating data from multiple and sometimes incompatible sources. This chapter provides a survey on the data mining techniques used frequently in web-based educational systems.

3.2 Clustering, Classification, Partitioning, Subgroup Discovery and Community Detection

Clustering is the process of grouping objects into classes of similar objects ([45]). It is an unsupervised classification or partationing of patterns into groups or subsets (clusters). This technique groups records together based on their locality and connectivity within an n-dimensional space. Clustering and classification are both classification methods ([50])

In eLearning area cluster analysis is used to identify students with similar behaviour [60], to determine students'learning style [115], [116] and to perform clustering on the learner answer [88].

The authors in [115] proposed to apply two-phase hierarchical clustering algorithm, to enable tutors to determine some parameters such as maximal number of groups, clustering treshold and weights for different learning style dimensions. In [115] is proposed the system architecture, in which teaching paths as well as proper layouts are adjusted to groups of students according to their learning styles and usability preferences, the author considered the usage of two versions of two-phase hierarchical clustering algorithm for students' grouping.

In [88] the authors considerated original set of questions of the e-questionnaires as input space, results into a huge number of unique features to be taken into consideration when performing clustering on the learner answers.

In [20] is presented a solution for advising the learner regarding the resources he should access and study in order to obtain a required proficiency level. The solution uses a Naive Bayes as classification algorithm with input data represented by learners performed activities.

3.2.1 Subgroup Discovery

Subgroup discovery represents a form of supervised inductive learning in which, given a set of data and a property of interest to the user, an attempt is made to lacate subgroups which are statistically "most interesting" for the user. The objective is to discover characteristics of subgroups with respect to a specific property of interest. In [71] authors describe the application of subgroup discovery using evolutionary algorithms and obtain rules which describe relations between the student's usage of the different activities and modules provided by the Moodle.

3.2.2 Tools for Clustering, Classification, Community Detection and Partitioning

- KEEL ([89]) - a software tool to assess evolutionary algorithms for Data Mining problems of various kinds including as regression, classification unsupervised learning, etc. In [10] authors used k-Nearest Neighbour classifier and XCS classifier from KEEL.
- WEKA ([90]) - a collection of Java implementations of Machine Learning algorithms (classification, regression, clustering, association rules, and visualization).
- ADaM - to be used in grid or cluster environments.
- RapidMiner ([91]) - learning schemes for regression, classification, and clustering tasks

3.3 *Sequential Pattern Mining*

Sequential pattern mining is the mining of frequently occuring ordered events or subsequences as patterns. It was first presented by Agrawal and Srikant [5], and based on finding inter-session patterns such as the presence of a set of items is followed by another item in a time-ordered set of sessions or episodes.

The main objective of Sequential Pattern Mining is to discover all frequent sequences of itemsets in a dataset. An itemset is a non-empty subset of elements from a set C, the item collection. A *sequence* is an ordered list of itemsets (also called only items). The number of items in a sequence is the length of the sequence: A sequence of a length k is called a k-sequence. The support for a sequence s is the number of sequences of which s is a subsequence noted $Sup(s)$. A subsequence has not be contiguous; formally, a sequence $a = \langle a_1, a_2, \ldots, a_n \rangle$ is a *subsequence* of $b = \langle b_1, b_2, \ldots, b_m \rangle$, if there exist integers $1 \leq i_1 < i_2 < \ldots < i_n \leq m$ such that $a_1 = bi_1, a_2 = bi_2, a_n = bi_n$.

Mining sequential patterns is a challenging process becuase such mining may test and generate a huge number of intermidiate subsequences.There are many approaches to acheive this task, Among them are the apriori-based [81] and the pattern-growth methods [40, 69].

Apriori-based methods encounter problems with mining of large sequence databases. Therefore the computation process takes long time. It can bear some nontrivial costs like potentially huge set of candidate sequences, multiple scans of databases or difficulties at mining long sequential patterns [81]. Acheiving efficient mining of sequential patterns was studied by many researchers [58, 39], abd this problem is usually solved either by using data reduction methods or pattern-growth methods.

Pattern-growth methods work with the complete set of patterns. The reduction of the computation intensity is applied in the subsequence generation phase. The PrefixSpan algorithm [69] is an example of this method, it reduces the size of the projected databases.

Recently both methods are used as the basis for other pattern mining algorithms. The usage of constraints was suggested to solve the problem mentioned above [12, 38]. A Constraint is defined as a predicate on the set of finite sequences. Then, with defined minimum support threshold s and a constraint we can claim that a sequence is frequent if it is contained in at least s sequences in the database and satisfies the constraint. Using constraints also reduces the search space, which contributes significantly to achieve better performance and scalability levels.

3.3.1 Sequential Pattern Mining in eLearning

Learning Management Systems (LMS) or Adaptive Hypermedia Systems (AHS) produce large amount of data stored in log files or databases. Data collections usually consist of various types of information. We can obtain typical information like event (activity, item), type of event, device or time when event was performed, identification of user (usually IP adress or login information) ...etc.

Sequential pattern mining methods usually are used for discovering which learning content has motivated access to other contents , discovering of common (frequently used) learning paths and student's learning behavior etc. Moreover, with combination of other computation intelligence methods, Sequential Pattern Mining is used to form user recommendation and adaptive systems, and to build tools that enable educators to recognize students' problems early.

Sequential pattern mining was applied by Zaiane and Luo [114] for discovering useful patterns based on restrictions to help evaluate student's activities in course. Pahl and Donnellan [68] analyzed a student's individual sessions calculating session statistics and searching for session patterns and time series. Usage of sequential pattern mining with defined constraints for acquiring background knowledge in Intelligent tutoring systems was presented in [11]. [48] was oriented to identification of significant sequences of student's activity. Finding of incorrect student behavior, finding of relationships between patterns of a learners behavior (including time spent online, number of read and published articles, number of asked questions etc.) is presented in [113]. In [109] were proposed several data mining techniques including sequential pattern mining to extract learning features and to predict which group a new learner belongs to. Sequential pattern mining was used for links recommendation in AHS in [66]. The authors developed a tool which was integrated to a well known AHA! system. Development of a mining and visualizing tool in order to help instructors to discover the most visited trails is presented in [72].

3.4 Association Rules Mining

3.4.1 Basic Concepts

Association rules were first introduced by Agrawal et al. in 1993 as means of determining relationships among a set of items in a database [3]. An association rule is an implication that one item (or a set of items) is associated with another item (or a set of items). Association rules mining can be defined as follows:

I = {i1, i2,..., im} is a set of items.
D = {t1, t2, ..., tn} is a set of transactions, called a transaction database, where each transaction t has a tid and a t-itemset.

An association rule is written as $X \rightarrow Y$,
meaning that whenever X appears Y also tends to appear. X and Y may be single items or sets of items (where $X \subset I$, $Y \subset I$, itemsets X and Y do not intersect). Item X is called the antecedent and item Y is called the consequence. Each association rule has two quality measurements, support and confidence. The support of an item is the percentage of transactions in which the item can be found, so the support of an association rule can be defined as the percentage of transactions in the database that contain both X and Y . The confidence for an association rule is the ratio of the number of transactions that contain both X and Y to the number of transactions that contain X. That is, support represents frequencies of occurring patterns; and confidence represents strength of implication.

3.4.2 Association Rules Advantages and Disadvantages

Association rules mining is one of the most well studied data mining techniques. Eventhough, they are used widely in learning management systems, there are also other specific problems related to their application to e-learning data. These drawbacks were outlined in [36], and the authors described some possible solutions for each one of them. In [110] the advantages and disadvantages of association rules are listed as follows:

Advantages of Association Rules:

- Association rules are readily understandable
- They are best suited for categorial data analysis
- The outcomes are easy to interpret and explain

Disadvantages of Association Rules:

- Association rules generate many rules, and sometimes these rules are trivial or poorly understandable.
- They are not expressions of cause and effect, they only describe relationships in particular datasets.
- Most of the current data mining tools are too complex for educators to use.

3.4.3 Association Rules Mining Algorithms

There are many association rule discovery algorithms, three of them are presented in [17]: Apriori, sampling and partitioning. The Apriori is the first and foremost among them (Agrawal et al. 1996) [4]. It is based on the property that any subset of a large itemset is also large. The sampling algorithm uses portions of the database, while the partitioning algorithm divides the database into partitions that can be mined individually. Most association rule mining algorithms require the user to set at least two thresholds, one of minimum support and the other of minimum confidence. Therefore, the user must possess a certain amount of expertise in order to find the right support and confidence settings to obtain the best rule. A really important improvement to the Apriori algorithm for use in educational environments is the Predictive Apriori [77] because it does not require the user to specify any of these parameters (either the minimum support threshold or confidence values). This algorithm aims to find the N best association rules, where N is a fixed number entered by the teacher. Another improved association rule mining algorithm is proposed in [112] , in this version data cube was introduced. The algorithm had two advantages, the execution time was short and the precision of the rules was high.

3.4.4 Applications of Association Rules Mining in eLearning

In web based education the learners are in different age level, gender, and social role, their culture, education background, attention and interest are also causes a great difference. Giving corresponding learning content developers to realize teaching learners according to their needs is very difficult [47]. Its basic reason lies in being difficult by obtaining the relations between the learner's personality characteristics and learning behavior patterns accurately, automatically. In this way, it is necessary to mine out the association rules between personality characteristics and learning behavior patterns. Association rules mining have been applied to web-based education systems extensively for building recommender agents that could provide advise for both students and teachers. In [37] a system is proposed oriented to find, share and suggest the most appropriate modifications to improve the effectiveness of the course. The authors apply association rule mining to discover interesting information through students usage data in the form of IF-THEN recommendation rules. They have also used a collaborative recommender system to share and score the recommendation rules obtained by teachers with similar profiles along with other experts in education. Adaptive Educational Hypermedia (AEH) Systems automatically guide and recommend teaching activities to every student according to their needs, these systems use association rules to discover patterns in the LMS data. In [107] the researchers demonstrate how data mining techniques can be used to discover and present relevant pedagogic knowledge to the teachers. Based on these techniques, a tool for teachers was developed to support the evaluation of adaptive courses. Another application of association rules mining was proposed in [24], the authors employ the association rules to mine the learners' profiles for diagnosing learners common learning misconceptions during the learning processes. This study applies the discovered association rules of the common learning misconceptions to tune the courseware structure through modifying its difficulty parameters in the courseware database, so that the learning pathway is appropriately tuned.

3.5 Soft Computing and E-Learning

The last decades have witnessed a growing interest for e-learning and expert systems. The growing is due to effect the Internet and communication technologies evolution on all aspects of our life. A knowledge-based subsystem can enhance the capabilities of decision support not only by providing the subject matter knowledge, but also by providing expertise in data management and modeling. All technologies use knowledge, which is organized in a knowledge base, to provide the needed support of e-learning, information search and retrieval. These are considered as applications of soft computing.

The idea of Soft Computing (SC) is still in its initial stage of crystallization. According to Lotfi Zadeh, who coined the term "Soft Computing is an emerging approach to computing which parallels the remarkable ability of human mind to reason and learn in an environment of uncertainty and imprecision". In other words, soft computing combines techniques taken from fuzzy logic, neural network, genetic algorithm, probabilistic reasoning and signal processing tools such as wavelet transform to obtain robust solutions at low cost for problems which would be intractable by conventional means. Soft Computing is a partner in its domain. The principal constituent methodologies in SC are complementary rather than competitive [64, 86, 87, 87]. Soft Computing is the basis of Computational Intelligent which encompasses swarm intelligence (SI), artificial immune systems (AIS). Soft computing usually refers a collective set of CI paradigms and probabilities methods [30, 29]. Using hybrid intelligent system is a promising research field of soft computing focusing on synergistic combination of multiple approaches to develop and solve the complexity problem of e-learning. For further details for these methods, we recommend [23, 74, 2]. Fuzzy logic is an important part of soft computing group. It plays a key role in the ways in which humans deal with complexity and imprecision. Most of previous studies of student's performance did not consider their behavior prediction. However, Felix et al have considered it and presented a framework using three methods of soft computing, namely, fuzzy inductive reasoning (FIR), rule extraction fuzzy inductive reasoning (LR-FIR) and causal relevance rule fuzzy inductive reasoning (CR-FIR). The obtained resulted showed enhancing of the system understanding and valuable knowledge to teachers about the course performance [21].

Juan et al, used aggregating evaluation by consumers through the use of fuzzy linguistic aggregation operators. In addition, they presented how the evaluation system can be fine-tuned based on well-known Artificial Intelligence techniques and fuzzy quantified queries [84].

In [33] proposed a neural network with fuzzy logic model which is called neuro-fuzzy to investigate inferring student characteristics and to find the optimal path of students based on their profiles such as capabilities, attitudes, knowledge level, motivation and intellectual ability. It allows making a knowledge-based decision making and the teacher can get a lot of benefit of this method.

To enhance the multimedia e-learning instead of using the log and text data, Ye et al applied their proposed method based fuzzy logic on extracting speech segments from a media resource which improved the accuracy of the speech detection, and enhances the efficiency of the voice activity detection (VAD), comparing with the single feature based VAD algorithm [85].

3.6 Artificial Neural Networks and Evolutionary Algorithms

Neural networks (NN), also called artificial neural networks (ANN) are used as a solution for many applications like classification, optimization, signal processing, and control. Likewise, they are often used for modeling of complex relationships between inputs and outputs or for finding significant patterns in data collections. NN attempt to simulate the structure and functional aspects of biological neural networks, so they can be classified as non-linear statistical data modeling tools.

Basic principle consists in building the model of interconnected neurons which are basic building blocks of the nervous system. The synapses of the neuron are modeled as weights. The strength of the connection between an input and a neuron is noted by the value of the weight[42]. The output of a NN relies on the cooperation of the individual neurons within the network. Main property of a neural network is that it can perform its overall function even if some of the neurons can not work. This is possible due to neuron's collaboration within the network [13].

ANN is an neural network that can change its structure based on external or internal information that flows through the network during the learning phase. Mathematical model is defined by a function $f : X \to Y$. The function $f(x)$ is defined as a composition of other functions $g_i(x)$, which can be defined as a composition of other functions. A widely used type of composition is the nonlinear weighted sum, which can be defined as:

$$f(x) = K\left(\sum_i w_i g_i(x)\right)$$

where K is predefined function and g is represented by a vector of functions $g = (g_1, g_2, \ldots, g_n)$. Activation function can be represented by trashold, Piecewise-linear function, or sigmoid function.

Dependencies between variables are represented with arrows in the network structure, widely used are multilayer neural networks. There are two basic approaches for NN topologies [42]:

- *Feed-forward neural networks*, where the data are processed from input to output neurons. The data processing can extend over multiple neurons (or layers of neurons), but no feedback connections are present. Typical example of feed-forward NN are Perceptron and Adaline or Madaline.
- *Recurrent neural networks* that do contain feedback connections. In this type of NN, the dynamical properties of the network are important. In the process some activation values of neurons are changed with relevance to dynamical behavior, some activation values evolve to a stable state when neurons are in relaxation process. Recurrent networks were published by well known researchers like Anderson, Kohonen or Hopfield, recent technologies are their modifications.

The main property of ANN is the ability to learn, which is shown with the improvement of its performance. It could be described as the acquisition of knowledge about its environment, during the iteration of the learning process. We can distinguish two basic types of learning process:

- *Supervised learning* or associative learning, when the network is trained by providing input and matching output patterns. The input/output information can be provided by an external subject (or system which contains the neural network). This type of learning can be provided through paradigms like Error correction learning, Reinforcement learning, or Stochastic learning. The main tasks solved withing this paradigm are pattern recognition (also known as classification) and regression (also known as function approximation). Well-known technique for training neural networks is usage of backpropagation algorithm. The supervised learning paradigm is also applicable to sequential data.
- *Unsupervised learning* or self-organization learning do not require any external element to adjust the weight of the communication links to their neurons. The system is supposed to discover statistically salient features of the input population. In general there two basic kinds of unsupervised learning: Hebbian learning

and Competitive learning. The applications include clustering, the estimation of statistical distributions, compression and filtering.
- *Reinforcement learning* can be considered as intermediate form of the above two types of learning. NN does some action on the environment and gets a feedback response from the environment. The environment is usually modeled as a Markov decision process (MDP). Common tasks solved within paradigm of reinforcement learning are control problems, games and other sequential decision making tasks.

Computational intelligence includes several methods, among which we must mention artificial intelligence. Researchers publish applications of evolutionary algorithm techniques like genetic and evolutionary programming. Genetic algorithms are often used in data mining and for pattern recognition[31]. There are the two basic approaches in pattern recognition using genetic algorithms:

- applying genetic algorithms as a classifier[14],
- using genetic algorithms for reseting or optimization of parameter's value in other classifiers like parameters in the classification process, feature selection[59] or for selection the types of classifiers [54].

Implementation of genetic algorithms consists in simulation of abstract representations (also called genomes) of candidate solutions (individuals or phenotypes) to an optimization problems. Evolution usually starts with a population of randomly generated individuals and happens in generations. Each generation has evaluated the fitness of every individual in the population. On the basis of fitness, individuals are stochastically selected from current population and modified (mutated) to form a new population (reproduction). This population is then used in next iteration of algorithm. When maximum of generations has been produced, or fitness level has been reached, genetic algorithm terminates.

3.6.1 Aplication in eLearning

Ribeiro and Cardoso proposed using neural network with support vector machine (SVM) to build prediction models able to track students behavior. The main object of SVM is to reform and reduce the data which will feed to neural network for training and testing. They used Moodle data collection from Course of Discrete structure of the informatics engineering bachelor at the university of Coimbra [70]. In [46] is described approach based on hidden Markov model for building student's behavior models. Authors used this approach in their learning by teaching environment where students learn by teaching a computer agent named Betty using a visual concept map representation. Bidgoli and Punch presented the approach for classifying students in order to predict their final grades [63]. For weighting the feature vectors in classification process authors used genetic algorithm to optimize the prediction accuracy. In [57] is proposed method which uses multiple feed-forward neural networks to dynamically predict students' final achievement and to cluster them in two virtual groups, according to their performance. Multiple-choice test grades were used as the input data set of the networks.

4 Multiagent Systems

Web based learning is an important element in modern teaching environment. Many Web based tools are becoming available and agent technology contributes substantially to this field. Agent technology has been applied to different eLearning applications such as: information retrieval, teaching , tutoring , assignment checking, adaptive learning ...etc. The authors in [67] proposed an emotional intelligent tutoring system based on multi-agent architecture to design an adaptive distributed collaborative and peer to peer e-learning environment. These agents manage both cognitive and affective model of the learner, and are able to recognize the learners facial expression through emotional agents and express emotions through emotional embodied conversational agents (EECA). Another adaptive eLearning system called CAWAS (Culturally AWAre System) was described in [18]. In this paper a tool was implemented to create cultural templates of multimedia documents that take in consideration the learners cultural specifications.

Multiagent recommendation systems are widely used in web-based education. In [83] a new approach for achieving adaptivity in education was described. This approach is based on the students' learning styles. while in [118]Personalized Recommendation Education Services (PeRES) was developed. The system consists of a number of agents related to each other hierarchically and collaborate with each other to provide the personalized study plan, search, recommendation and communication services for learners and instructors and achieve the system overall goal-Personalized Education and re-use of courseware.

The dropout rate in e-learning is higher than that in traditional learning due to its low degree of continuity. researchers found that this dropout rate is related to the degree of learner's satisfaction. To analyze the learner satisfaction a webbased e-learning system based on multiagents and neural networks has been developed in [25].The system constructs the learners satisfaction network model about the e-learning grouping. Based on this network model, the proposed system can decide if the group remains, or is reorganized,or breaks down.

5 eTutors

In eLearning systems instructors/tutors/mentors undertake a vital teaching role that differs from that of a traditional classroom teacher. It is frequently asserted that the use of tutors is a major factor in achieving high student satisfaction, and low drop-out rates. Online tutors support e-learners through different roles [16, 62], these roles can be classified as:

- pedagogical roles: tutors support the learning process itself by providing instructions, stimulating questions, examples, feedback, motivation etc. to the learners
- managerial roles: The managerial role requires the tutor to perform basic course administration, track student progress and data etc.
- social roles: the tutor's social role includes the efforts to establish a friendly and comfortable environment and a community that stimulates learning

- technical roles: the technical role requires the tutor to acquaint the students and himself/herself with the ICT that is used for e-learning, and also to provide some technical support to the students

To be successful in all the required roles, an online tutor has to possess certain skills. Thomas made a 4P checklist according to which a good online tutor should be: positive, proactive, patient and persistent (in Shepherd, 2002). Many authors emphasize the need for tutors to be experts in the field they are tutoring. Tutors should also have highly developed online communication skills, be ICT literate and familiar with the e-learning technology and have a positive attitude toward students and learning [79].

One of eLearning systems' purposes is to encourage knowledge sharing so that valuable knowledge embedded in the network can be effectively explored. Most of the learners participate in this type of education with the expectations that they can acquire and share valuable knowledge to fulfill their needs.

Wachter et al. pointed out that an enhanced learning environment is possible only if one goes beyond online course delivery and creates a community of learners and other related resource groups [108, 111]. Wasko and Faraj found that knowledge sharing has been a motivation for participation in virtual communities [61, 111].

In Collaborative Learning the students are engaged in an open-ended effort to advance their collective understanding. They are encouraged to rely on each other as sources of information and assistance. In addition, interactions among the students facilitate learning directly by encouraging them to explain the subject matter to each other and revealing in a constructive way the inconsistencies and limitations in their knowledge. This participation which takes place in a meaningful social context enables a group of the students to acquire the skills of the eTutor and play his roles when he/she is not available. However, in some eLearning systems it is difficult to achieve efficient and effective knowledge sharing due to the following two barriers: (1) the difficulty in finding quality knowledge, and (2) the difficulty in finding trustworthy learning collaborators to interact with.

An important difference between user-generated content and traditional content is the variance in the quality of the content. In social media the distribution of quality has high variance: from very high-quality items to low-quality; this makes the tasks of filtering and ranking in such systems more complex than in other domains. In that case, models of credibility which are used extensively on search engine research and information retrieval can be used in order to evaluate the trustworthiness of the students' knowledge.

Several graph theoretic models of credibility rely strongly on the consideration of the indegree of the node (the sum of the incoming arcs of a node in a directed graph) so as to extract importance and trustworthiness. However, there are social activities (e.g. collaborative authoring) which derive much of their credibility by their productions (e.g. authorship). In that case, the in-degree cannot provide input to evaluate the importance of that entity and therefore an alternative evaluation is needed, which has to consider the outputs of the entity (productions) [51].

PageRank and HITS were the pioneering approaches that introduced Link Analysis Ranking, in which hyperlink structures are used to determine the relative authority of a Web page. PageRank assumes that a node transfers its PageRank values evenly to all the nodes it connects to. A node has high rank if the sum of the ranks of its in-links is high. This covers both the case where a node has many in-links and that where a node has a few highly ranked in-links. This can be clarified by the following example given in c [19, 117]. If B is able to answer A's questions, and C is able to answer B's questions, then C should receive a high authority score, since he is able to answer the questions of someone who himself has some expertise. PageRank provides interesting results when the interactions between users are around one specific subject only. The study in [117] illustrates this by using a PageRank-like algorithm called ExpertiseRank on data from the Java forum, in which the interactions between users are exclusively about Java programming.

The fundamental assumption of HITS is that in a graph there are special nodes that act as hubs. Hubs contain collections of links to authorities (nodes that contain good information). A good hub is a node that points to good authorities, while a good authority is a node pointed to by good hubs [49]. So, askers can act as hubs and best answerers can act as authorities. HITS associates two scores to each node: a hub score and an authority score.

6 Multilingual E-Learning

With the advent of Internet, English language was the lingua franca of Web resources, and search engines. Consequently, internet English speakers have reaped significant benefits from Internet and search engines than non-English speakers. Currently, the World Wide Web has appeared as an indispensable and dynamic resource of information in almost all major languages which make the Internet a truly multilingual world. The Internet is the most important information resource that

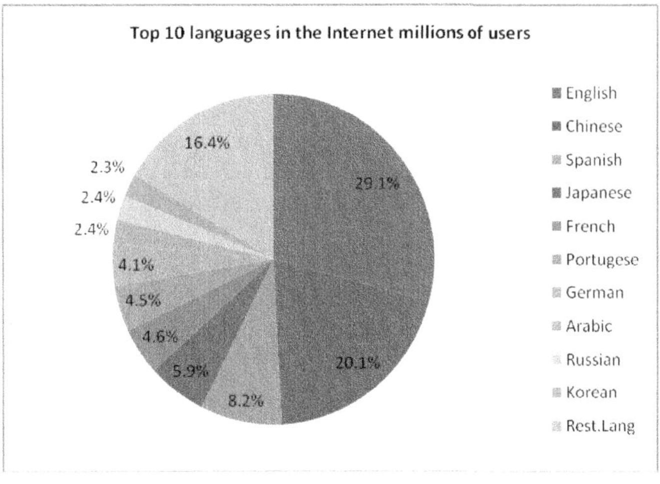

Fig. 1. Top 10 Languages in the Internet

contains a large repository of multilingual documents and web sites widespread on cyberspace. A report published by Internet World Stats in 2009 showed that 65% of Internet users are from non-English speaking areas as shown in figure 1.

Nowadays, Internet has shown new paths to learning with the evolution of the traditional Web into the WEB 2.0. Nevertheless the unequal use of the language and the users' lack to access the relevant educational materials in a language that they speak and understand seem to reinforce the growing knowledge gap between information and cognition processing on the Internet. This problem has been motivating UNESCO to consider one of its important goals, "to achieve worldwide access to e-contents in all languages, improve the linguistic capabilities of users and create and develop tools for Internet multilingual access".

6.1 Multilingual e-Learning Tools

Multilingual e-learning tools use language resources to help users figure out the best learning material that they can get to satisfy their needs. Abe et al [1] addressed a design and development of multilingual e-learning system of 17 different languages including European and Asian languages. The system is called TUFS language Education module system [92]. They used two different teaching methods called "Audio-Lingual method" and "Verbo-Tonal method". In [43], Hrad et al showed project ELefANTS (E-Learning for Acquiring New Types of Skills) that was concluded in 2007. The main objective of the project is to help the Internet users to enter the labor market and increase their opportunities to be integrated into the work process. The project provided courses of seven languages (English, French, German, etc). Verdú et al [106], described the ODISEAME project as an effort to extend the use of the Internet and technologies, services and applications related to it, to the European Union and some middle east countries. Sharing and transfer technology and knowledge, with the aim of reducing the existing barriers for digital inclusion were one of the main objectives of this project.

Denev et al included Bulgarian language within three open sources, namely Moodle [93], LAMS [94] and WeLOAD [95]. This work is step toward of multilingual e-learning system. They concluded that there is no universal approach for creating multilingual software and advised to use Modular software architecture with export into external scripts "strings" structured in language packs independent of the program logic. [27].

Some researchers tried to create a new domain of research by dealing with the languages and the documents as signal and using a mathematical transformation for processing. This new domain is useful and promising to facility the multilingual information access which is a critical problem for several applications, especially to multilingual e-learning and multilingual search. In [7, 6], Al-Dubaee and Ahmad applied a new direction for wavelet transform on multilingual Web information retrieval of 14 languages belonging to 5 language families. Forty-one wavelet functions (mother wavelets) of six families, namely, Haar (haar), Daubechies (db2-10),

Biorthogonal (bior1.1- 6.8 and rbior1.1 - 6.8), Cofilets (coif 1-5), Symlets (sym 1 - 10), and Dmey (dmey) are investigated in order to evaluate a suitable wavelet function for multilingual web information retrieval. Their novel method is based on the sentence query entered by Internet user that is converted to signal using its Unicode standard (unique identifier code). Unicode is an international standard for representing the characters used into plurality of languages. Also, it provides a unique numeric character, regardless of language, platform, and program. In [8], the multiwavelet transform has proved the ability for being a multilingual web information retrieval, regardless of type, script, word order, direction of writing and difficult font problems of languages study. It has solved the problems of selecting optimum wavelet transform for sentence query entered of previous 14 languages of five language families by Internet user. In [64, 86, 87, 87], Mitra et al have considered a wavelet transform as a new tool of signal processing in Soft Computing (SC). The multiwavelet is a body of wavelet transform. Due to this fact, they suggested to consider multiwavelet in Soft Computing as well. In [9], the aptitudes of multiwavelet transform to represent one language (domain) of the multilingual and multicultural world regardless of type, script, word order, direction of writing and difficult font problems of the language was further investigated with more languages and families. As a result of their work with 31 sentence queries for 29 languages belong to 8 language families, they expect that the wavelet or multiwavelet transform becomes multilingualism tool on the Internet. They advised it as a universal approach for creating multilingual software.

6.1.1 European Union Funded Multilingual e-Learning Projects

- LT4eL (Language Technology for e-Learning) is a European Union project to improve the use of multilingual language technology tools and semantic web techniques for enhancing the retrieval and the metadata annotation of learning material [96]. The researchers of this project have developed new functionalities such as a key word extractor and a glossary candidate detector, tuned for eight languages (Bulgarian, Czech, English, ...etc) by using the tools and resources into ILIAS [52, 80, 65, 28]. (ILIAS is an open source LMS) [97].
- WebALT (Web Advance Learning Technology) is a project established during the period 2005 to 2007 is as Web service to develop multilingual contents for an XML databases of mathematical problems to be used in undergraduate university courses in mathematics. It enables the natural language generation of the mathematics text into seven languages by using the MathML, Maple A.P, Moodle and OpenMath softwares. It aims to improve the access to Europe's educational resources in mathematics and provide new forms of learning experience. WebALT has produced a variety of software tools such as:
  - WebALT grammars to enable the natural language generation of the mathmatics text into the required languages.
  - TextMathEditor is used for authoring mathematical text conforming to the WebALT grammars.

  - WExED is an WebALT Exercise Editor and supporting the creation of language independent problems in mathematics.
  - WebALT Maple T.A. Firefox plug-in is enhancing Maple T.A Whereas it provides question content in multilingual features. WabALT is now a commercial under Webalt company [98].

- eColoRe (eContent localization Resources for Translator Training) project provides shareable materials to support localization of eContent and computer-assisted translation (CAT) training. This project commenced in November 2002 and is finished in 2005.
- MeLLANGE (Multilingual e-learning in LANGuage Engineering) project extended to eColoRe which was completed in April 2005. It is also the European Union funded project which launched from 1 October 2004 to 20 September 2007. The aim of this project is to create innovative learning materials for trainee and professional translators by using academic and industry groups from eight language countries.
- E2ngineering is a project to improve development and testing of multilingual e-learning materials and courses in advanced engineering subjects based on reusable elements (LOMs), second generation e-learning architecture and methodology. It uses five languages to supply multilingual e-learning access to information [99].
- Bony is a project of multilingual and cognitive e-learning management system which allows users to find and to learn and access to materials via personal digital assistant (PDA) phones to achieve their training experience. At the same time, it is a social network for users whereas it provides a social network experts fully dedicated to the European cooperation in the domains of research and project management [100]. The essential objectives of this project are to supply the labour market needs with the realization of an "intelligent and multimedia e-courses" for eleven different languages and to enhance foreign languages acquisition in combination to vocational training with the support of multilingual semantic technologies and to improve training accessibility and mobility.

6.1.2 Commercial Software and Tools Providing Multilingual e-Learning

- Basis Technology is a commercial company which provides software solutions for extracting meaningful intelligence from unstructured multilingual text in Asian, European and Middle Eastern languages. It helps technology companies and government organizations (Cisco, Microsoft, Oracle, Symantec, Google, Yahoo etc.) in several spheres such as the accuracy of information retrieval, text mining and other applications through advanced linguistics [101].
- Trivantis is multilingual e-learning commercial tools which consists of two tools, namely CourseMill 5.0 and Lectora 2008 [102]. CourseMill 5.0 and Lectora 2008 are learning management systems (LMS) that help companies distribute increased sales opportunities and they provide five languages(Dutch, French, German, Italian, and Spanish) and four languages (French, German, Spanish and Swedish) respectively.

7 Case Study: Moodle System

This case study is oriented to find latent social networks from a number of data collections based on students' similar behavior. These data collections are stored in Moodle system logs used at Silesian University, Czech Republic. These logs consist of records of all events performed by Moodles users such as communication in forums and chats, reading study materials or blogs, taking tests or quizes etc. The users of this system (students, teachers, and administrators) are members of a community with one target - to achieve the learning process successfully. The authors of this chapter are interested in studying students' activities and discovering the latent social network among this group of students.

7.1 Data Preprocessing

The data related to this case study was exported from a relational database to a number of data logs (log files in a text format). Moodle logs record all events performed by all users (students, tutors and administrators). In this experiment only the events performed by students were analyzed, events performed by tutors and administrators were removed. Student' records were represented by lines in the text file, and each record has a number of attributes, see table (1).

Table 1. Attributes of one record

Attribute Name	Description
NameOfCourse	Course name
TimeStamp	Start date and time of event (in national format)
IPAdress	IP address of computer student accessed the course from
NameOfStudent	Student full name identified by his/her login
NameOfEvent	Event performed by student in LMS (e.g. resource view, blog view, quiz attempt)
IDOfEvent	Event ID represents detailed information of event (e.g. number of resource view)

The data preprocessing phase consists of the following steps:

- Data anonymousness - the attribute NameOfStudent was replaced by attribute StudentNumber. This attribute was used for record identification, the IPAddres attribute was ignored.

- Event definition and set of events creation:
  - Definition of event - an event was represented by a combination of attributes: NameOfEvent_IDOfEvent_NameOfCourse. Each event was identified by internal EventID.
    Example: quiz view_1288_OPF-ZS-07/08-SV/ESOC-E;0 - internal EventID is = 0.
  - Creation of sets of events - consisted of events for each student in a certain course ordered by TimeStamp. Example: NameOfCourse;StudentNumber; TimeStamp;EventID_1;EventID_2; ... ; EventID_m.
- Activity processing:
  - Activity definition - activity was described by sequence of events. Each activity obtained its ActivityID: EventID_1;EventID_2; ... ; EventID_m - ActivityID.
    Example: 0;1;2;3;4; - 1.
  - Creation of sets of activities - for each student we obtained a set of activities identified by ActivityID, for each activity was obtained ActivityAmount: ActitivyID ActivityAmount.
    Example: Student_1:
    0 2
    3 1
    2 1

Furthermore, additional data corrections were implemented. Students who executed only one activity were removed from the data set. This type of activity misrepresented the intensity of a relationship. Many of such students logged into the Moodle system only once and have not pursued the study activities. This behavior is typical for distance education, where a large number of students usually do not finish the studies (for various reasons typical for this type of education).

Every event was defined by two attributes: NameOfEvent and and IDOfEvent (e.g. NameOfEvent = resource view, IDOfEvent = 1203). This combination was obtained for events like resource view, forum view, user view etc. Every event in LMS Moodle is represented by its number. For example the resource view number represents the unique reading material. However, there are events like forum view forum, resource view resource, user view all, without an ID. These events represent the first step to perform a certain event (e.g. if student would like read some material, he/she must first open list of resources (NameOfEvent = resource view all, IDOfEvent = ""), and after that goes to the material. Since they was not relevant to this work, events without an ID were removed too. This removal reduced both the number of events per students and the number of activities per student, but the graph inclination practically did not change. The reduction did not influence the finding of optimal time period.

We have found that from time period 15 minutes long the number of student activities practically does not change. Due to this fact, it was not necessary for us to pass through every event in sequence stored in the whole log.

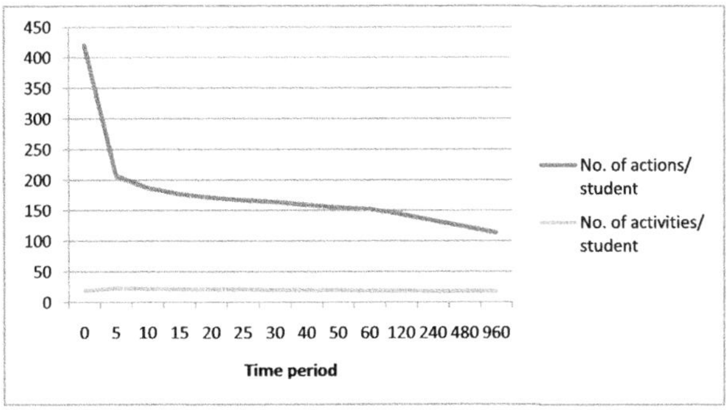

Fig. 2. Graph of time period

In consequence, the activities were created by event sequences of a 30 minute time periods. In our experiments described above, 30 minutes time periods have been proven to be the most effective time interval. The findings showed that in shorter time periods (5 minutes) students were performing only non-study activities, and in longer periods there was not a significant activity difference (that means activity types were very similar), see figure 2. Similar conclusion was presented by Zorrilla et al. in [119].

For each student we obtained a set of activities executed in the Moodle system. For comparing the number of events and activities per student in relation to time period, see table 2.

Table 2. Number of Activities and Events Performed by Students in Different Time Periods (where ES is No. of events/student, AcS is No. of activities/student, AllAc is No. of all activities, AllE is No. of all events).

Time period	ES	AcS	AllAc	AllA
1	420	19	251	12445
5	207	23	646	14587
10	187	22	633	13952
15	177	21	615	13565
20	171	21	610	13289
25	167	21	602	13120
30	164	20	601	12954
40	159	20	588	12659
50	155	20	580	12481
60	152	19	579	12382
120	143	19	563	12016
240	133	18	544	11793
480	123	18	537	11678
960	113	18	536	11621

7.2 *Similarity between Students and Visualization*

Two matrices were used to represent the data: the StudentMatrix A (Students X Activity types) and the MatrixOfSimilarity S (Students X Students), which is derived from matrix A, and defines students' relationships using their similar activities. The similarity between two students (vectors of sets of activities) was defined by the Cosine measure [76].

$$S_{i,j} = \frac{\sum_{k=1}^{n} a_{ik} a_{jk}}{\sqrt{\sum_{k=1}^{n} a_{ik}^2} \sqrt{\sum_{k=1}^{n} a_{jk}^2}} \tag{1}$$

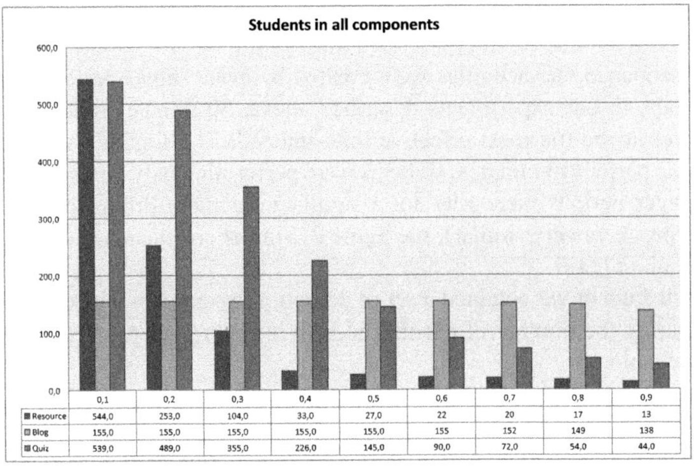

Fig. 3. Graph of Number of Students at Different Levels of Similarity

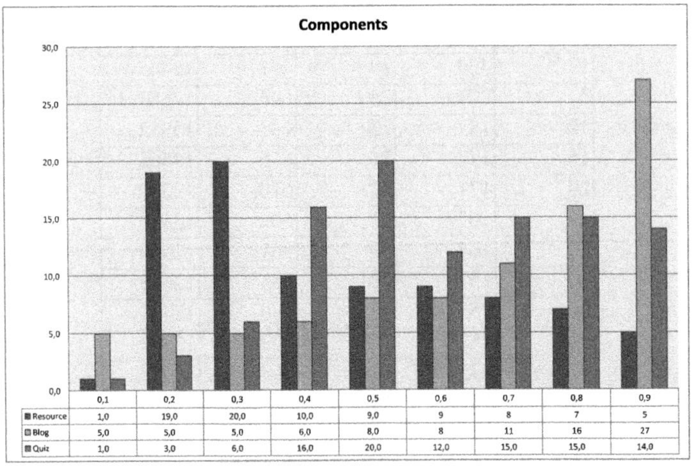

Fig. 4. Graph of Number of Components at Different Levels of Similarity

Fig. 5. Visualization of SN - Resource with similarity level = 0.2

Fig. 6. Visualization of SN - Quiz with similarity level = 0.2

The results of this experiment were represented by the following graphs, the data was divided into three groups (quiz, blog and resources). For each group a number of weighted graphs were created with different levels of similarity.

Figures 3 and 4 demonstrate the dependency between the activities and the size of components which consist of students with similar learning behavior.

In graph 3, number of students in all components for similarity levels from 0.1 to 0.9 we can see that number of students in activities like "quiz" and "resources" is decreasing with hyperbolic function, while in the "blog" activities the number of students is stable. In fig 4 the number of components in the "resources" and the "quiz" also depends on the level of similarity, while the "blog" is again stable - there are strong relations between components in blog. For more details, see figures 5, 6 and 7 below.

Graphs 5,6 and 7 are the visual representation of the social network, where the nodes represent the students and lines represent the relationships between them, with a 0.2 similarity level.

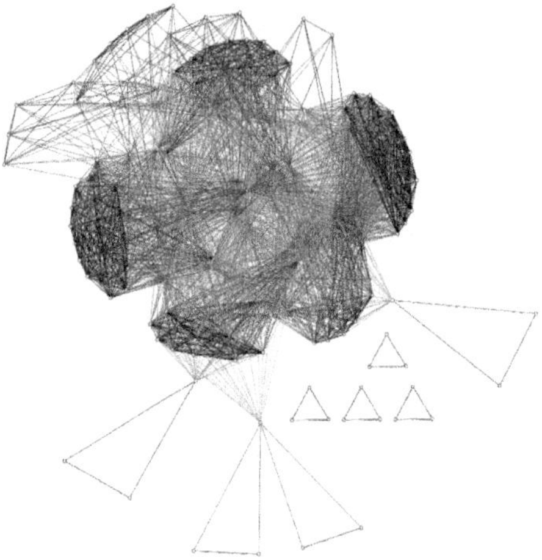

Fig. 7. Visualization of SN - Blog with similarity level = 0.2

8 Conclusion

In this chapter a survey about soft computing and data mining techniques and their application in eLearning was introduced, followed by a case study of students' activities in an eLearning system. The results of the survey showed that there are still some challenges and complex tasks to be solved, and there is a great demand to find some tools and alternatives to solve these tasks that require computing power and storage capacity. These problems will be studied in our future work, we plan to study some advanced techniques and algorithms related to voice and image identification which might provide good opportunities for students with special needs. Besides we are interested in studying the implementation of grid computation in eLearning to achieve the sharing and re-usability of resources among the different schools and in recommended systems to simplify the learning process on the basis of learners' behavior.

References

1. Abe, S., Nakata, S., Kigoshi, T., Mochizuki, H.: Designing and developing multilingual e-learning materials: Tufs language education pronunciation module - introduction of a system for learning japanese language pronunciation. In: ICALT, p. 462 (2003)
2. Abonyi, J., Balasz, F., Abraham, A.: Computational intelligence in data mining. Informatica 29, 3–12 (2005)
3. Agrawal, R., Imielinski, T., Swami, A.N.: Mining association rules between sets of items in large databases. In: SIGMOD Conference, pp. 207–216 (1993)
4. Agrawal, R., Mannila, H., Srikant, R., Toivonen, H., Verkamo, A.I.: Fast discovery of association rules. In: Advances in Knowledge Discovery and Data Mining, pp. 307–328. AAAI/MIT Press (1996)

5. Agrawal, R., Srikant, R.: Mining sequential patterns. In: Proceedings of International Conference on Data Engineering (ICDE 1995), pp. 3–14. IEEE Computer Society Press, Los Alamitos (1995)
6. Al-Dubaee, S.A., Ahmad, N.: The bior 3.1 wavelet transform in multilingual web information retrieval. In: DMIN, pp. 707–713 (2008)
7. Al-Dubaee, S.A., Ahmad, N.: New direction of applied wavelet transform in multilingual web information retrieval. In: FSKD, (4), pp. 198–202 (2008)
8. Al-Dubaee, S.A., Ahmad, N.: A novel multilingual web information retrieval method using multiwavelet transform. UbiCC 3(3) (2008)
9. Al-Dubaee, S.A., Snášel, V., Ahmad, N.: Wavelet, multiwavelet and multilingualism on the internet. In: IKE (2009)
10. Alcalá-Fdez, J., Sánchez, L., García, S., del Jesus, M.J., Ventura, S., Garrell, J.M., Otero, J., Romero, C., Bacardit, J., Rivas, V.M., Fernández, J.C., Herrera, F.: Keel: a software tool to assess evolutionary algorithms for data mining problems. Soft Comput. 13(3), 307–318 (2008)
11. Antunes, C.: Acquiring background knowledge for intelligent tutoring systems. In: Proceedings of Educational Data Mining 2008: 1st International Conference on Educational Data Mining, pp. 18–27 (2008)
12. Antunes, C.M., Oliveira, A.L.: Inference of Sequential Association Rules Guided by Context-Free Grammars. In: Adriaans, P.W., Fernau, H., van Zaanen, M. (eds.) ICGI 2002. LNCS (LNAI), vol. 2484, pp. 289–293. Springer, Heidelberg (2002)
13. Arbib, M.A. (ed.): The Handbook of Brain Theory and Neural Networks. The MIT Press, Cambridge (2002)
14. Bandyopadhyay, S., Muthy, C.A.: Pattern clasification using genetic algorithms. Pattern Recognition Letters 16, 801–808 (1995)
15. Benôit, G.: Data mining. Annual Review of Information Science and Technology 36, 265–310 (2002)
16. Berge, Z.L.: Facilitating computer conferencing: Recommendations from the field. Educational Technology 35(1), 22–30 (1995)
17. Berry, M.W., Browne, M.: Lecture notes in data mining, ch. 14. World Scientific, Singapore (2006)
18. Blanchard, E., Razaki, R., Frasson, C.: Cross-cultural adaptation of elearning contents: a methodology. In: International Conference on E-Learning (2005)
19. Bouguessa, M., Dumoulin, B., Wang, S.: Identifying authoritative actors in question-answering forums: the case of yahoo! answers. In: KDD 2008: Proceeding of the 14th ACM SIGKDD international conference on Knowledge discovery and data mining, pp. 866–874. ACM, New York (2008)
20. Burdescu, D.D., Mihaescu, M.C., Logofatu, B.: Employing bayes classifier for improving learner's proficiency. In: International Conference on Internet and Web Applications and Services, pp. 38–42 (2008), `http://doi.ieeecomputersociety.org/10.1109/ICIW.2008.16`
21. Castro, F., Nebot, A., Mugica, F.: A soft computing decision support framework to improve the e-learning experience. In: SpringSim, pp. 781–788 (2008)
22. Castro, F., Vellido, A., Nebot, A.: Minguillon: Finding relevant features to characterize student behavior on an e-learning system. In: Hamid, R.A. (ed.) FECS 2005, pp. 210–216 (2005)
23. Castro, F., Vellido, A., Nebot, A., Mugica, F.: Applying data mining techniques to e-learning problems. Studies in Computational Intelligence(SCI), vol. 62 (2007)
24. Chen, C.M., Hsieh, Y.L., Hsu, S.H.: Mining learner profile utilizing association rule for web-based learning diagnosis. Expert Syst. Appl. 33(1), 6–22 (2007)

25. Cho, Y.I.: Intelligent automatic community grouping system by multiagents. Artificial Life and Robotics 12(1-2), 284–290 (2008)
26. Daradoumis, T., Martinez-Mones, A., Xhafa, F.: A layered framework for evaluating on-line collaborative learning interactions. International Journal of Human-Computer Studies 64(7), 622–635 (2006)
27. Denev, D., Boichev, A., Nedelchev, P.: Multilingual support of e-learning systems: The bulgarian contribution. In: CompSysTech 2006 (2006)
28. Dolog, P., Gavriloaie, R., Nejdl, W., Brase, J.: Integrating adaptive hypermedia techniques and open RDF-based environments. In: WWW - Alternate Paper Tracks (2003)
29. Eberhart, R.C., Shi, Y.: Computational Intelligence: An Introduction. Morgan Kaufmann, San Francisco (2007)
30. Engelbrecht, A.P.: Computational Intelligence: An Introduction. J. Wiley & Sons Inc., Chichester (2007)
31. Falkenauer, A.A.: A survey of Evolutionary Algorithms for data Mining and Knowledge Discovery. In: Ghosh, A., Tsutsui, S. (eds.), pp. 819–845. Springer, New York (2003)
32. Fayyad, U., Piatetsky-Sshapiro, G., Smyth, P.: From data mining to knowledge discovery in databases. AI Magazine 17, 37–54 (1996)
33. Fazlollahtabarand, H., Mahdavi, I.: User/tutor optimal learning path in e-learning using comprehensive neuro-fuzzy approach. Education Research Review Elsevier 4(2), 142–155 (2009)
34. Fernández, C.A., López, J.V., Montero, F., Gonzalez, P.: Adaptive interaction multi-agent systems in e-learning/e-teaching on the web. In: Cueva Lovelle, J.M., Rodríguez, B.M.G., Gayo, J.E.L., Ruiz, M.d.P.P., Aguilar, L.J. (eds.) ICWE 2003. LNCS, vol. 2722, pp. 144–153. Springer, Heidelberg (2003)
35. Frawley, W.J., Piatetsky-Shapiro, G., Matheus, C.J.: Knowledge discovery in databases: An overview. In: Knowledge Discovery in Databases, pp. 1–30. AAAI/MIT Press (1991)
36. García, E., Romero, C., Ventura, S., Calders, T.: Drawbacks and solutions of applying association rule mining in learning management systems. In: Proceedings of the International Workshop on Applying Data Mining in e-Learning 2007. IEEE Press, Los Alamitos (2007)
37. García, E., Romero, C., Ventura, S., Castro, C.D.: An architecture for making recommendations to courseware authors using association rule mining and collaborative filtering. User Modeling and User-Adapted Interaction 19(1-2), 99–132 (2009)
38. Garofalakis, M., Rastogi, R., Shim, K.: Mining sequential patterns with regular expression constraints. IEEE Transactions on Knowledge and Data Engineering 14, 530–552 (2002)
39. Han, J., Dong, G., Yin, Y.: Efficient mining of partial periodic patterns in time series database. In: ICDE 1999: Proceedings of the 15th International Conference on Data Engineering, p. 106. IEEE Computer Society, Washington (1999)
40. Han, J., Pei, J., Mortazavi-Asl, B., Chen, Q., Dayal, U., Hsu, M.C.: Freespan: Frequent pattern-projected sequential pattern mining. In: Proceedings of the 6th International Conference on Knowledge Discovery and Data Mining (KDD 2000), pp. 355–359. ACM, New York (2000)
41. Hand, D.J., Mannila, H., Smyth, P.: Principles of Data Mining. In: Adaptive Computation and Machine Learning. The MIT Press, Cambridge (2001), `http://www.amazon.ca/exec/obidos/redirect?tag=citeulike09-20%&path=ASIN/026208290X`

42. Haykin, S.: Neural Networks: A Comprehensive Foundation, 2nd edn., p. 842. IEEE Press, Los Alamitos (1998)
43. Hrad, J., Zeman, T., Hájek, J.: Multilanguage e-learning course for industrial automation. In: Learning to Live in the Knowledge Society, pp. 371–372 (2008)
44. Hwang, G.J., Huang, T.C.K., Tseng, C.R.: A group-desigion approach for evaluating educational web sites. Computers & Education 42, 65–86 (2004)
45. Jain, A.K., Murty, M.N., Flynn, P.J.: Data clustering: a review. ACM Computing Survey 31(3), 264–323 (1999)
46. Jeong, H., Biswas, G.: Mining student behavior models in learning-by-teaching environments. In: Baker, R.S.J.d., Barnes, T., Beck, J.E. (eds.) Proceeings of Educational Data Mining 2008: 1st International Conference on Educational Data Mining, pp. 127–136 (2008)
47. Jun, L., Renhou, L., Qinhua, Z.: Study on the personality mining method for learners in network learning. Journal of Xian Jiaotong University, 575–576 (2004)
48. Kay, J., Masionneuve, N., Yacef, K., Zaïane, O.: Mining patterns of events in students' teamwork data. In: Proceedings of the Workshop on Educational Data Mining at the 8th International Conference on Intelligent Tutoring Systems (ITS 2006), pp. 45–52 (2006)
49. Kleinberg, J.M.: Authoritative sources in a hyperlinked environment. J. ACM 46(5), 604–632 (1999)
50. Klösgen, W., Zytkow, J.M. (eds.): Handbook of data mining and knowledge discovery. Oxford University Press, Inc., New York (2002)
51. Korfiatis, N., Poulos, M., Bokos, G.: Evaluating authoritative sources using social networks: An insight from wikipedia. Online Information Review 30(3), 252–262 (2006), `http://www.korfiatis.info/papers/OISJournal_final.pdf`
52. Kubon, V., Spousta, M.: Multilingual approach to e-learning from a monolingual perspective. In: FLAIRS Conference, pp. 229–230 (2008)
53. Kumar, A.: Rule-based adaptive problem generation in programming tutors and its evaluation. In: The 12th International Conference on Artificial Intelligence in Education, pp. 36–44 (2005)
54. Kuncheva, L.I., Jain, L.C.: Designing classifier fusion systems by genetic algorithms. IEEE Transaction on Evolutionary Computation 33, 351–373 (2000)
55. Lee, S.H., Kim, P.J., Ahn, Y.Y., Jeong, H.: Googling hidden interactions: Web search engine based weighted network construction (2007), `http://www.citebase.org/abstract?id=oai:arXiv.org:0710.3268`
56. Li, L., Otsuka, S., Kitsuregawa, M.: Query recommendation using large-scale web access logs and web page archive. In: Bhowmick, S.S., Küng, J., Wagner, R. (eds.) DEXA 2008. LNCS, vol. 5181, pp. 134–141. Springer, Heidelberg (2008)
57. Lykourentzou, I., Giannoukos, I., Mpardis, G., Nikolopoulos, V., Loumos, V.: Early and dynamic student achievement prediction in e-learning courses using neural networks. Journal of the American Society for Information Science and Technology 60(2), 372–380 (2009)
58. Mannila, H., Toivonen, H., Verkamo, A.I.: Discovery of frequent episodes in event sequences. Data Mining of Knowledge Discovery 1(3), 259–289 (1997)
59. Martin-Bautista, M.J., Vila, M.A.: A survey of genetic feature selection in mining issues. In: Proceeding Congress on Evolutionary Computation (CEC 1999), pp. 1314–1321 (1999)
60. Martinovič, J., Dráždilová, P., Slaninová, K., Snášel, V.: Relation analysis in elearning. In: International Conference on Computer Information Systems and Industrial Management Applications, pp. 133–138. IEEE Computer Society, Los Alamitos (2008)

61. McLure-Wasko, M., Faraj, S.: Why should i share? examining social capital and knowledge contribution in electronic networks. MIS Quarterly 29(1), 35–57 (2005)
62. McPherson, M., Nunes, M.B.: The role of tutors as an integral part of online learning support (2009), `http://www.eurodl.org/materials/contrib/2004/Maggie_MsP.html`
63. Minaei-Bidgoli, B., Punch, W.F.: Using Genetic Algorithms for Data Mining Optimization in an Educational Web-Based System. In: Cantú-Paz, E., Foster, J.A., Deb, K., Davis, L., Roy, R., O'Reilly, U.-M., Beyer, H.-G., Kendall, G., Wilson, S.W., Harman, M., Wegener, J., Dasgupta, D., Potter, M.A., Schultz, A., Dowsland, K.A., Jonoska, N., Miller, J., Standish, R.K. (eds.) GECCO 2003. LNCS, vol. 2724, Springer, Heidelberg (2003)
64. Mitra, S., Acharya, T.: Data Mining Multimedia, Soft Computing, and Bioinformatics. J. Wiley & Sons Inc., Chichester (2003)
65. Monachesi, P., Lemnitzer, L., Simov, K.I.: Language technology for elearning. In: EC-TEL, pp. 667–672 (2006)
66. Morales, C.R., Pérez, A.R.P., Soto, S.V., Martínez, C.H., Gómez, A.Z.: Using sequential pattern mining for links recommendation in adaptive hypermedia educational systems. In: Proceedings of 5th international Conference on Multimedia and Information and Communication Technologies in Education, M-CITE 2006 (2006)
67. Neji, M., Ben, A.M.: Agent-based collaborative affective e-learning framework. The Electronic Journal of e-Learning 5(2), 123–134 (2007)
68. Pahl, C., Donnellan, C.: Data mining technology for the evaluation of web-based teaching and learning systems. In: Proceedings of the congress e- learning (2003)
69. Pei, J., Han, J., Mortazavi-asl, B., Pinto, H., Chen, Q., Dayal, U., Hsu, M.-c.: Prefixspan: Mining sequential patterns efficiently by prefix-projected pattern growth. In: Proceedings of 17th International Conference on Data Engineering (ICDE 2001), pp. 215–224 (2001)
70. Ribeiro, B., Cardoso, A.: Behavior pattern mining during the evaluation phase in an e-learning course. In: Proceeings of International Conference on Engineering and Education (2007)
71. Romero, C., González, P., Ventura, S., del Jesus, M.J., Herrera, F.: Evolutionary algorithms for subgroup discovery in e-learning: A practical application using moodle data. Expert Syst. Appl. 36(2), 1632–1644 (2009)
72. Romero, C., Gutiérrez, S., Freire, M., Ventura, S.: Mining and visualizing visited trails in web-based educational systems. In: Proceedings of Educational Data Mining 2008: 1st International Conference on Educational Data Mining (2008)
73. Romero, C., Ventura, S. (eds.): Data Mining in E-Learning. WIT Press, Southhampton (2006)
74. Romero, C., Ventura, S.: Data Mining in E-learning. WIT Press, Southampton (2006)
75. Romero, C., Ventura, S., Garcia, E.: Data mining in course management systems: Moodle case study and tutorial. Computers & Education 51(1), 368–384 (2008), `http://dx.doi.org/10.1016/j.compedu.2007.05.016`
76. Salton, G., Buckley, C.: Term-weighting approaches in automatic text retrieval. In: Information Processing and Management, pp. 513–523 (1988)
77. Scheffer, T.: Finding association rules that trade support optimally against confidence. Intell. Data Anal. 9(4), 381–395 (2005)

78. Schommer, C.: An unified definition of data mining (2008), http://www.citebase.org/abstract?id=oai:arXiv.org:0809.2696, http://arxiv.org/PS_cache/arxiv/pdf/0809/0809.2696v1.pdf
79. Shepherd, C.: In search of the perfect e-tutor (2009), http://www.fastrak-consulting.co.uk/tactix/features/perfect_etutor.htm
80. Simov, K.I., Osenova, P.: Bulqa: Bulgarian-bulgarian question answering at clef 2005. In: Peters, C., Gey, F.C., Gonzalo, J., Müller, H., Jones, G.J.F., Kluck, M., Magnini, B., de Rijke, M., Giampiccolo, D. (eds.) CLEF 2005. LNCS, vol. 4022, pp. 517–526. Springer, Heidelberg (2006)
81. Srikant, R., Agrawal, R.: Mining sequential patterns: Generalizations and performance improvements. In: Apers, P.M.G., Bouzeghoub, M., Gardarin, G. (eds.) EDBT 1996. LNCS, vol. 1057. Springer, Heidelberg (1996)
82. Stathacopoulou, G.D., Grigoriadou, M.: Neural network-based fuzzy modeling of the student in intelligent tutoring systems. In: International Joint Conference on Neural Networks, pp. 3517–3521 (1999)
83. Sun, S., Joy, M., Griffiths, N.: The use of learning objects and learning styles in a multi-agent education system. In: Proceedings of the World Conference on Educational Multimedia, Hypermedia and Telecommunications, AACE, pp. 3403–3410 (2005)
84. Tao, Y., Zu, D., Li, X., Du, P.: A fuzzy aggregation-based reputation model for e-learning exploitation of public domain resources. In: 4th International ICSC Symposia on Soft-Computing and Intelligent Systems for Industry (2001)
85. Tao, Y., Zu, D., Li, X., Du, P.: A fuzzy logic based speech extraction approach for e-learning content production. In: ICALIP 2008, pp. 298–302 (2009)
86. Thuillard, M.: Wavelet in Soft Computing. Word Scientific series in Robotics and intelligent systems, Singapore, vol. 25 (2001)
87. Thuillard, M.: Adaptive multiresolution and wavelet based search methods. International Journal of Intelligent Systems 19, 303–313 (2004)
88. Tzouveli, P., Mylonas, P., Kollias, S.: An intelligent e-learning system based on learner profiling and learning resources adaptation. Comput. Educ. 51(1), 224–238 (2008)
89. http://www.keel.es/ (2008)
90. http://www.cs.waikato.ac.nz/~ml/weka/ (2008)
91. http://rapid-i.com/ (2008)
92. http://www.tufs.ac.jp/21coe/language (2008)
93. http://www.moodle.org (2008)
94. http://www.lamsfoundation.org (2008)
95. http://weload.lernnetz.de (2008)
96. http://www.lt4el.eu/ (2008)
97. http://www.ilias.de/ (2008)
98. http://www.webalt.net/ (2008)
99. http://www.e-collegium.org/e2ngineering/index.html (2008)
100. http://www.bonynetwork.eu (2008)
101. http://www.basistech.com/ (2008)
102. http://www.personneltoday.com/articles/2008/04/29/45581/trivantis-multilingual-elearning-tools-now-available.html (2008)
103. http://www.google.com (2009)
104. http://www.google.com/trends (2009)
105. http://sp.ask.com/en/docs/iq/iq.shtml (2009)

106. Verdú, E., Verdú, M.J., Regueras, L.M., de Castro, J.P.: Intercultural and multilingual e-learning to bridge the digital divide. In: Human.Society@Internet, pp. 260–269 (2005)
107. Vialardi, C., Bravo, J., Ortigosa, A.: Improving aeh courses through log analysis. J. UCS 14(17), 2777–2798 (2008)
108. Wachter, R.M., Gupta, J.N.D., Quaddus, M.A.: It takes a village: Virtual communities in supporting of education. International Journal of Information Management 20(6), 473–489 (2000)
109. Wang, W., Weng, J.F., Su, J.M., Tseng, S.S.: Learning portfolio analysis and mining in scorm compliant environment. In: 34th Annual Frontiers in Education, FIE 2004, vol. 1, pp. 17–24 (2004)
110. Wickramasinghe, N., Gupta, J.N.D., Sharma, S.K.: Creating knowledge-based healthcare organizations, ch. 7. Idea Group (2004)
111. Yang, S.J.H., Chen, I.Y.L., Kinshuk, Chen, N.S.: Enhancing the quality of e-learning in virtual learning communities by finding quality learning content and trustworthy collaborators. Educational Technology & Society 10(2), 12 (2007)
112. Yin, C.Y., Luo, Q.: Personality mining system in e-learning by using improved association rules. In: 2007 International Conference on Machine Learning and Cybernetics, vol. 7, pp. 4130–4134 (2007), doi:10.1109/ICMLC.2007.4370869
113. Yu, P., Own, C., Lin, L.: On learning behavior analysis of web based interactive environment. In: Proceedings of ICCEE (2001)
114. Zaiane, O., Luo, J.: Web usage mining for a better web-based learning environment. In: Proceedings of conference on advanced technology for education, pp. 60–64 (2001)
115. Zakrzewska, D.: Cluster analysis for users' modeling in intelligent e-learning systems. In: Nguyen, N.T., Borzemski, L., Grzech, A., Ali, M. (eds.) IEA/AIE 2008. LNCS (LNAI), vol. 5027, pp. 209–214. Springer, Heidelberg (2008)
116. Zakrzewska, D.: Using clustering technique for students' grouping in intelligent e-learning systems. In: Holzinger, A. (ed.) USAB 2008. LNCS, vol. 5298, pp. 403–410. Springer, Heidelberg (2008)
117. Zhang, J., Ackerman, M.S., Adamic, L.: Expertise networks in online communities: structure and algorithms. In: WWW 2007: Proceedings of the 16th international conference on World Wide Web, pp. 221–230. ACM Press, New York (2007), `http://portal.acm.org/citation.cfm?id=1242603`
118. Zhu, F., Ip, H.H.-S., Fok, A.W.P., Cao, J.: Peres: A personalized recommendation education system based on multi-agents & scorm. In: Leung, H., Li, F., Lau, R., Li, Q. (eds.) ICWL 2007. LNCS, vol. 4823, pp. 31–42. Springer, Heidelberg (2008)
119. Zorrilla, M., Menasalvas, E., Marín, D., Mora, E., Segovia, J.: Web Usage Mining Project for Improving Web-Based Learning Sites. In: Moreno Díaz, R., Pichler, F., Quesada Arencibia, A. (eds.) EUROCAST 2005. LNCS, vol. 3643, pp. 205–210. Springer, Heidelberg (2005)

Advanced Learning Technology Systems in Mathematics Education

Valeria Marina Monetti[1], Loredana Randazzo[1], Antonello Santini[3], and Gerardo Toraldo[1,2]

[1] Department of Agricultural Engineering, University of Naples Federico II, Via Università 100, 80055, Portici - Napoli, Italy
[2] High Performance Computing and Networking Institute Italian National Research Council Via P. Castellino, 111, 80131, Napoli, Italy
[3] Department of Food Science, University of Naples Federico II, Via Università 100, 80055, Portici - Napoli, Italy

Abstract. Mathematics education is a very active field of interest, with many journals and books focusing on the large variety of topics involved in such process, whose growing importance is mainly due to the needs of professional skills in the use of effective computational and modelling tools in solving real world problems. In this work we deal with the impact of technologically advanced learning technologies on mathematics education; we also focus on how web based approaches might suggest new paradigms of mathematical learning aimed to face in an effective way the new educational challenges of the modern advanced ICT and Network Society.

1 Introduction

Web based learning tools aimed to make effective the online education process implementation have been strongly developed in the last ten years, finding wide diffusion in High Schools as well as in graduate and undergraduate University classes. The main goal is to benefit of educational materials through a web user interface by adopting, very often, a cognitive model close to the one laying beyond traditional courses.

According to Rekkedal and Qvist-Eriksen (Rekkedal and Qvist-Eriksen, 2003), who move from the definition of distance education in (Keegan, 1996), online education is a form of education characterized by:

1. the quasi-permanent separation of teacher and learner throughout the length of the learning process (this distinguishes it from conventional face-to-face education);
2. the influence of an educational organization both in the planning and preparation of learning materials and in the provision of student support services (this distinguishes it from private study and teach yourself programmes);
3. the use of computers and computer networks to unite teacher and learners and carry the content of the course;

F. Xhafa et al. (Eds.): Computational Intelligence for Tech. Enhanced Learning, SCI 273, pp. 225–248.
springerlink.com

4. the provision of two-way communication via computer networks so that the student may benefit from or even initiate dialogue (this distinguishes it from other uses of technology in education);
5. the possibility, non mandatory, to use online learning groups during the learning process so that individuals rather than as groups are considered, even if online meetings, either face-to-face or mediated by electronic tools, for both didactic and socialization purposes can be scheduled and arranged.

Usually an educational process is based on a blend of learning activities using traditional and technologically advanced supports. In distance learning, the role of the Information and Communication Technologies (ICT) is crucial; and therefore the continuous development of ICT makes e-learning a rapidly evolving field. E-learning is just part of the more general process of creation of the so called "Network Society" in which the demand for "anytime-anywhere education" becomes stronger and stronger.

Although a number of innovative projects are rapidly evolving, from the methodological point of view there are still many gaps to be filled in among the different components involved in an efficient educational process. Despite the huge amount of educational material available online (lecture notes, interactive presentations, handbooks, teaching software, computational tools, etc.), the goal of merging all these components in a unifying online educational framework is still far from accomplished. In particular, to create interactive Learning Environments (LE) based on ICT the use of automatic system is highly recommended (Moallem, 2001; Marchioro and Landon, 1997; Pohjolainen et al., 2003; Smith and Ragan, 1993; Harasim, 1996).

The Learning Technology Systems (LTS) (Avgeriou et al., 2001; Rafaeli et al. 2004; Kong and Kwok, 1998; Sheremetov and Arenas, 2002) that provides knowledge through the internet and web technologies is a powerful tool to integrate knowledge and build interdisciplinary educational paths (Michailidou and Economides, 2003): "They often provide facilities to adapt to the context allowing users to change certain system parameters and adapt their behaviour accordingly" (Oppermann et al.,1997).

The impact of web learning in mathematics education is even more dramatic than in other disciplines, since technologically advanced approaches potentially allow integration of basic statistical-mathematical knowledge in a computationally effective framework.

The technologically advanced and knowledge-intensive world asks for qualified professional skills in the effective use of technological and computational tools with requirements from a simple data representation to the development of complex systems scenarios through a process of modelling and synthesis of the real world. In other words, online teachers and students are required to have a scientific and technical background that allows the full exploitation of the potentialities of the widely available Information and Communication Technologies (ICT) resources. These potentialities dramatically increased in the last ten years, due to the availability of powerful and cheap computers and, at the same time, of user-friendly software tools (interactive computational environments), allowing users to easily access large data set and computing facilities.

An up to date e-learning system includes basically three components:

1. System Management Tools. These are a variety of tools or accessories for creating web pages or sites, including all functionalities needed for user-management such as registration and access privileges to the platform of users according to their roles, authentication procedures, student tracking, curriculum management. They also include the creation and management of the databases with all of the contents including analysis tools and search engines.
2. Communication and Assessment Tools These are the tools needed for accessing the contents, and organising and facilitating the collaborative creation and exchange of material to foster and support the user community.
3. Learning Materials.

The first component is the most technical one that is less dependent than the other components on pedagogical issues and on academic disciplines; it is based on software tools which are now well established. The second and third component require a fine tuning with the academic discipline at hand. This paper will examine System management Tools, Communication and Assessment Tools, and Learning materials. There will be a short overview of projects, issues, trends, theoretical and practical technologies in modern teaching, and learning approaches in the math education field through the extensive use of technologically advanced ICTs.

2 General Remarks on the Assessed Criteria for Teaching Using a Web Oriented Platform

When dealing with e-learning and asynchronous learning systems in distance education, there are many concerns and issues to consider with e-learning platforms and web environments. When first entering a virtual environment/class students do not enter a face-to-face classroom where they receive multiple verbal and non-verbal clues that help them gain a feeling for the atmosphere of the class. When students enter a web-based class, they are entering a non-physical, computer-mediated communication context, a workspace. Some of the main factors influencing atmosphere in an online workspace are look and feel, organization of content, and structure of interaction (Graziadei et al., 1997; Salmon 2000).

Unlike other online educational environments where instructors can customize background or menu colors, layout, and/or icons, the majority of the web environment is characterized by an highly standardized shell (Driscoll 2000). This represents a major benefit for students in that once they have taken one web-based course, all subsequent online courses will be immediately familiar to them. On the other hand, the instructors are not required to spend a lot of time remodelling the classroom every time they teach a course.

2.1 Organization of Content

How should the teacher determine where to place information? The course content area, for example, is essential, and it should be used to present introductory material. The heart of a web based online class should be the conference area, the main space where class activity and interaction take place. This area is highly versatile and can accommodate the most formally structured or spontaneously associative discussions. It can be used to present web-resources and directly link the reading of these resources to the discussion of them. Most contact takes place in this conference area. The use of this space and the role the students can play here require some attention: the teacher can decide if they can co-moderate discussions, lead discussions, and respond to questions and comments.

2.2 Interactions in a Computer Mediated Context: The Workspace

The instructor biography is one of the first areas students come in touch with. It is usually a dedicated space in the web learning platform environment containing a biography, an image, and a URL. This section that is essential to build the learning environment is available to the students when they are allowed into the space to browse prior to the beginning of any activity. This is a useful tool to familiarize the students with the virtual classroom environment. Students should also be encouraged to introduce themselves with a short biographical note. There are main areas of the workspace, where the teaching, knowledge transfer and learning activity take place. Some may be thought of as performing administrative tasks e.g. assignments, class members etc., while others may be viewed as places to perform academic tasks like conferences, study groups etc. The work space itself may be seen as static area, where storing information is done, or a dynamic area, that allows providing room for the learning activities. One of the most important tasks in course planning is finding a way to make full use of work space while creating a coherent learning environment (Pahl et al., 2009).

The areas univocally considered relevant in an online synchronous-asynchronous efficient workspace for advanced teaching are:

2.3 Syllabus, Course Introduction, and Content

A web course syllabus consists of multiple parts. It is important that the sections form a coherent whole, and that information within these sections moves progressively from beginning to end. Web distance education platforms should use the syllabus area to present course objectives and perspectives. However, a more general problem to consider before the syllabus assignment is the syllabus in relation to the general task of disseminating introductory course information in the asynchronous, online workspace. The sections of the syllabus should point to areas such as description of general content, projects and assignments, exercises, online testing, grading.

While all introductory material is important, the heart of it is in the general course introduction or statement of purpose. This information is crucial in providing the students with a feeling for the subject matter and the method of the work ahead. Course content is a ready reference and resource area put together specifically for the course needs. It often contains pre-designed course material, and it is very important to integrate the rest of the interface (conferences, study groups, readings, webliography, web available resources, etc.) into this pre-designed material. The organization of the course content involves many important pedagogical issues regarding questions to be addressed such as: what information should be placed into the course content; will the concepts for the type of material stored be apparent to the students; and will the course content give a clear picture of how this area will work together with the other areas. The content area should also include a location for students to submit their assignments and for teachers to assign grades and to place comments in students folders.

2.4 Conferences and Study Groups

Usually this is the most dynamically interactive section of a web learning interface. It can be used for question and answer sessions, discussions of all types, mini-lectures or student presentations or projects. The conference area is the heart of an efficient web teaching-learning interface: the place where most of the teaching is done. This is the teacher's regular contact point with the students, and it is where ideas can be exchanged, guidance can be given, questions can be asked, policies can be reviewed, and the structure of the learning strategy can be established. From this section links to web resources can be assigned or mini lectures presented as student response and reflection is collected. Study questions related to materials or textbook can be posed, and field questions about readings or documents downloaded or retrieved from the web and examined. These type of conferences are usually the core and the center of activity and energy of a web distance education interface.

The study group is an interesting and versatile resource that can be used for projects, collaborative brain storming, exchanging research resources, discussing ideas, and participating in peer editing activities or any number of other activities. This area allows to identify small groups of students and have them work intensively on precise projects. The use of study groups is, however, a controversial area of web teaching platforms since some students dislike the logistics of working in groups, and some teachers do not see the benefits of isolating groups from the rest of the class.

2.5 Chat Rooms and Video Conference

The use of these resources is useful for facilitating real time communication between all the subjects, students, and teachers involved in the virtual class, and it also facilitates interaction between the students and the exchange of information and materials or data. Different teachers use the chat room differently and also different students have different feelings about it, but in distance education, this is the synchronous area of instruction.

2.6 Webliography

This is generally considered an important feature of any dynamic distance education activity. Important URLs are stored here and should be integrated into the class activity schedule and directly related to the ideas of the course allowing the students to link to web sites considered relevant and explore ideas, concepts etc. Placing an annotated webliography of readings in the course content area with live links, and then using the webliography for student-recommended sites and/or sites that are included in weekly discussions is useful to instructors and students. The webliography, if the students are allowed to suggest new links or new materials or learning resources, can be considered a dynamic area of the work space where, after a peer review procedure, new materials and web resources can integrate the existing ones indicated in the course content links section.

The aforementioned needs or criteria developed for classifying activities in a the web workspace should be used with the goal of achieving a balance in a distance education online web-based class (Nichols 2008; Siemens 2003). Some of the criteria used for the selection of web based class activities assessment are: (1) reading and responding to selected web-based texts; (2) independent web research on special or related topics projects using web-based information; (3) textbook-oriented activities, e.g. textual analysis, definition, application of concepts, etc.; (4) library-oriented activities, e.g. research of hard copy resources or library databases; (5) self-contained activities like discussion of ideas presented within the conference, exchange of personal experiences, etc.

A few questions to be answered before to start off with the class, should be: the kind of activity needed for the best use of the available teaching material; a balance of discussion and presentation, formal and informal thoughts; student led and instructor led conversation; a balance between textbook, library resources or web-based resources. The objective of the workspace should be to provide a place for brainstorming and exchange questions and problems, creating new activities for the online work. To reach this goal, many online web platforms use the study groups as a powerful tool that gives a great impulse to the distance education teacher/student interaction and communication.

3 Dealing with the Language of Mathematics

The well established symbolism in mathematics has often been considered a unifying language in applied science that is able to foster dialogue and communication across the borders of disciplines and cultures. In a Physics book, for instance, the verbal description of a problem is usually strongly language dependent, although its mathematical formulation can easily be understood by people from many different nations. However the very specific mathematics language presents serious difficulties in making an effective web-learning system; a major drawback is that the many content management tools and web learning technologies have been initially developed to deal mostly with natural language. On the other hand, the universality of the mathematical language, once technical impediments about content rendering and exchange have been removed, might lead to the creation of effective multilingual learning environments for mathematics.

A mathematical content mark up language and formula representation should be globally accessible, flexible, and easily integrated within a popular text editor in a high-quality typesetting system, that is able to build technical and scientific documents without the constraints to use plain text or figures for mathematical formulas. This would allow to build learning material for the students (lecture notes, articles, exercises, questions with answer in closed form, etc.) in a correct ad homogeneous format, easy to be downloaded and upgraded. The features just outlined actually represent the minimal requests to be considered: an effective representation of mathematical contents is just a somehow minor issue that must be considered in a mathematics web-learning system. The most critical issues have to do with the assessment phase, and, more generally, with all the features of the e-learning platform that require high levels of interactivity and content update, exchange, and creation. Such features clearly denote the difference between a simple repository of on line-lectures and a web based learning system that uses a collaborative and interactive approach.

The use on the web of interactive mathematics, and even just the use of mathematical notation, has been a very hard task for a long time and this represented a very serious drawback on the use of advanced learning technologies in mathematics education. The way of using symbols or editors for rendering mathematical formulas was one of the first problems to deal with: such a task has been faced long before the birth of the web and effectively solved by the Donald E. Knuth's TeX typesetting system (Knuth, 1984); nowadays LaTeX[1], based on TeX, is the most classical markup language in mathematics. LaTeX is a document writing system for high-quality typesetting, that can be used for almost any form of publishing, although its features make it especially suitable for technical or scientific documents, where mathematics plays a meaningful role. Besides the typesetting of complex mathematical formulas, LaTex supports advanced documents publishing features such as the use of different sheet styles, the control on large documents containing sectioning, cross-references, tables and figures and the automatic generation of bibliographies and indexes.

The main drawback for the original LaTex is that it is a non-visual editor that requires the knowledge of a simple syntax and of editing efforts that are paid back by the excellent results that can be achieved. Moreover, several powerful ASCII editors have been made available in the last years that are easy-to-use environments for LaTex that allow to prepare nice looking scientific documents or presentations even when the author has very little knowledge or experience with the LaTex syntax. WinEdt[2], widely used as a front-end for HTML, is also an effective and popular front-end for compilers for different typesetting systems such as TeX that is integrated into a very powerful system with spell checking functionalities, multi-lingual setups, and dictionaries.

The strong limitations for math contents representation in HTML, which is the standard mark-up language on the web, has been overcome by the development of the Mathematical Mark-up Language (MathML)[3], the first standard language to be approved by the World Wide Web Consortium (W3C). MathML was first released as a W3C recommendation in 1998 according to the recommendations of the W3C with the specific aim to deal with mathematical expressions, and it was built on

[1] http://www.latex-project.org/

[2] www.winedt.com/

[3] http://www.w3.org/TR/MathML2/chapter2.html#fund_overview

the top of the Extensible Markup Language (XML)[4], delivered the same year as the standard format for representing structured information on the web. Nowadays, mathematical contents produced by MathML can be directly displayed by the most common web browsers. MathML is aimed to produce mathematical and scientific contents on the Web with an high level of flexibility. However, its features are not just limited to the presentation aspects: they involve semantic aspects through "Content MathML"[5] and therefore much more than just displaying formulas can be done. A simple formula editor is aimed to put formulas in human readable form but the same formulas might be displayed in different ways semantically equivalent although expressed as different objects. For automatic computer processing or even for a simple search procedure in web pages, they turn out to be ambiguous with no possibility to be used by any procedure which needs a univocal interpretation. A semantic content mark up such as Content MathML provides a semantically rich representation of the knowledge, suitable for logical processing and exchange of the information.

An obvious drawback of MathML is that it appears much less readable than TeX, and looks quite unsuitable to be used by humans since writing and editing are assumed to be done by automatic systems. For these reasons several converters from the most popular languages have been recently made available; for instance, LaTeXMathML[6] is capable of transforming a pure latex coding a "web-friendly" string of commands and converts the standard LaTeX mathematical syntax into MathML. Very similar are Java Library SnuggleTeX[7] and Blahtex[8] which is a program written in C++, designed for this purpose. Such programs convert from a "low semantic" to an "higher semantic" language, and this is a very challenging task because of the unavoidable ambiguity of a system such TeX, whose aim is to present contents with no attention to the meaning of the represented objects. Because of that, actually, just specific subsets of TeX (or Latex) are converted from any of the above mentioned programs. Converters to MathML are also available from others mark up languages different from TeX, like ASCIIMathML[9] which translates ASCII math text to MathML and it is very useful in translating Wiki-like texts[10].

A further step towards a more powerful semantically rich Mathematics mark up language is OpenMath, whose developers were aimed to encode mathematical objects that can be:

- displayed in a browser;
- exchanged between software systems;
- cut and pasted for use in different contexts;
- verified as being mathematically sound or not;

---

[4] www.w3.org/XML/

[5] W3C MATH WORKING GROUP (2003) Mathematical Markup Language (MathML) 2.0, 21 October 2003, http://www.w3.org/TR/MathML2/

[6] math.etsu.edu/LaTeXMathML

[7] http://www.ph.ed.ac.uk/snuggletex/

[8] http://gva.noekeon.org/blahtexml

[9] http://asciimathml.sourceforge.net

[10] http://www1.chapman.edu/~jipsen/mathml/asciimath.html

- used to make interactive documents really interactive;
- standard for representing the semantics of mathematical objects.

With respect to MathML, Openmath is "semantic-oriented" much more than "presentation-oriented". The two products share a strong inspiration from the recommendation formulated by the Worldwide Web Consortium; see [http://www.openmath.org/overview/om-mml.html] and reference therein for a comparison among the two. However has been observed that: *"The original motivation for OpenMath came from the Computer Algebra community. Computer Algebra packages were getting bigger and more unwieldy, and it seemed reasonable to adopt a generic "plug and play" architecture to allow specialised programs to be used from general purpose environments. There were plenty of mechanisms for connecting software components together, but no common format for representing the underlying data objects. It quickly became clear that any standard had to be vendor-neutral and that objects encoded in OpenMath should not be too verbose"*[11].

In other words, the primary goal is to deal with automatic computer processing, by supplying consistent input for computational software and any kind of mathematical manipulation; moreover a strong motivation is also the creation of databases of mathematical objects equipped with the classical functionalities such as indexing, retrieval, search.

Recently, user friendly formula editors that produce mathematics objects for automatic computer processing have been developed, with the aim to help in building semantic mathematical formulas in a user friendly framework, making more widely available the potentialities of products like OpenMath for educational purposes and their use in web learning platforms. An example could be the WIRIS[12] OM tools. WIRIS is a semantic oriented formula editor based on OpenMath, whose features have been inherited with the addition of graphical and functional extra features such as error checking in real time and multilingual tools. Open-Math is also equipped with an on-line Computer Algebra System (WIRIS CAS) and therefore is strongly aimed to integrate mathematics teaching and computing.

In conclusion, the enormous evolution of advanced semantic oriented web editors in the last few years allow the creation of powerful web-leaning environments for mathematics, supporting a semantically rich creation of contents and strongly integrating their exchange with computational environments, this way fulfilling also the semantic web aspirations. This is a typical case where technological advancement are designed to drive meaningful changes in educational paradigms and approaches.

4 Available Online Platforms and Educational Running Projects in Mathematics

There are numerous web platforms and educational web sites available on the Internet and the great part of them have been developed by Universities. In general

[11] http://www.openmath.org/

[12] http://www.wiris.com/

some of them are educational platforms that can host non specifically oriented discipline contents and that can hence be adapted to different courses and to different subjects, while others are specifically oriented, and this last choice is taking over recently as the election choice for advanced learning systems. A final remark should be to always consider in the evaluation of a web platform, its flexibility to specifically suite the students and course needs.

Here in the following are listed some platforms considered relevant that are used for web learning and online synchronous and asynchronous activities, some specifically designed projects that produced software or teaching/learning objects, and websites available in the web that host learning activities in the field of the Mathematics education:

Fermat: e-learning of Mathematics	www.fermat.uma.es
WebAlt	www.webalt.com
MathAssess	http://mathassess.ecs.soton.ac.uk/
SeLma	http://selmacity.schoolinsites.com/
ActiveMat	http://www.activemath.org
Fermat: e-learning of Mathematics	www.fermat.uma.es
EMILeA-stat	http://www.emilea.de
Webmath	http://www.webmath.com

In the following the relevant aspects of each platform and the learning technology adopted are briefly outlined.

4.1 *Fermat: e-Learning of Mathematics*

The main aim of this project, initiated in 2003, was originally to develop a web site to complement the education in Mathematics. It was initiated at the University of Malaga, in the Department of Applied Mathematics. The core of the platform was year by year developed using management content systems like Mambo[13] and Joomla[14], open source free available software very flexible for designing web pages. The platform introduces theoretical concepts and laboratory sessions where the students can work on real problems using software as MatLab[15] available and structured within the platform itself. The major peculiarities of the platform are: easy access to didactic documents in the web, interactive tutorials complementing the formation and allowing each student to verify his/her knowledge, availability of Forums that cover the main lecture topics (equivalent to the above mentioned "conference areas"), connection to University libraries, availability of a virtual tour of the School of Engineering of Telecommunication, multi language support. The core of the platform, that is mainly Mathematics oriented, is the possibility to use in an interactive mode the software like MatLab in the practical laboratory sessions designed for the students. The web platform also includes sections dedicated to information e.g. congresses, scientific articles of spreading on newspapers, news etc. Each course presented, in general, includes a presentation of the

[13] http://mambo-foundation.org/
[14] www.joomla.org/
[15] www.mathworks.com/

subject, relations of problems, interactive tests, tutorials, links and references on the subject with a strong connection to the local University libraries and facilities integrated in the web platform. When finalizing each course, surveys of quality are made that allow to detect problems, and take into account suggestions from the students. This web platform includes many of the abovementioned requirements considered as necessary for an online educational activity. It is however limited the communication between the main actors, teachers and students, of the learning process, being only realized using mainly asynchronous activities: a Forum area and e-mail communication.

4.2 WebAlt Project

This project combines existing standards for representing mathematics and existing linguistic technologies to create a language independent mathematical content for multi lingual and multi cultural localization. Differently from other projects, WebAlt is oriented to create a content that includes the specifically created WebAlt grammar for mathematics with a Multilanguage support that makes it extremely appealing to use in different languages that are automatically generated. WebAlt authoring tools and content support the Bologna process making it possible for example that different educational institutes in Europe to use the same test banks. The main features of this project, that has set up as a Company that commercializes premium contents and services to European educational Institutions, are the availability of a grammar for the language independent encoding of mathematics exercises, an editor, TextMathEditor, for authoring mathematical text in different languages, an e-repository, called Walter, for the storage of educational materials with an expanding metadata definitions, the WebAlt exercise editor, WexEd, that allows to create language independent problems in mathematics, an online mathematics laboratory for calculus and a portal to access all the content and materials of the project. The main features of the project are the TextMathEditor and the Multilingual Generator that can be used via online web service but the project offers also a free software solution, WexEd, the exercise editor and player of multilingual problem tree exercises represented using MathDox, that can be packaged in Sharable Content Object Reference Model (SCORM)[16] and then used as a third party environment like Moodle[17], taking advantage of the administrative facilities and grade-book provided in the learning environment. The software tools are available trough the project portal to registered authors that also can access the online repository, Walter, containing electronic resources for mathematics. Currently the repository stores many interactive exercises that can be downloaded in different formats and in a choice of languages. It is worth to note that a number of pilot studies in synchronous and asynchronous teaching in mathematics using the content produced by the project have been conducted and many High Schools and Universities in Europe offer online courses in mathematics using the WebAlt content. The e-learning system includes some of the abovementioned tools that are considered a general requirements for an online educational activity. Somehow it

---

[16] http://www.scormsoft.com/scorm

[17] http://www.moodle.org/

seems limited the communication between the main actors of the learning process, teachers and students, Forum area or better a conference area is missing and e-mail communication reduces the possible activities mainly to asynchronous activities. The possibility however to use the interactive section dedicated to exercises and the multi language interface make it appealing for online teaching activities.

4.3 MathAssess

Partner institutions on this project, Liverpool John Moores University and the Universities of Glasgow and Portsmouth, have used e-assessment as a crucial part of the delivery of mathematics courses. MathAssess replaces the system which they currently use, CALMAT, no longer supported, with a new platform derived from the previous one and that provides alternative facilities of comparable scope and flexibility. MathAssess utilizes a version of the Question and Test Interoperability specification (QTI) toolkit (ASDEL, AQuRate, Minibix, R2Q2)[18], extended to include the MathQTI specification in item authoring and rendering to handle mathematical expressions. To enable the use of algebraic expressions within questions, a Computer Algebra System is incorporated into the rendering and template processing calculations. The MathAssess project aims to build on the significant investment the Joint Information Systems Committee (JISC)[19] has made in QTI as an open standard in the e-assessment arena and supported the development of open source tool kits such as ASDEL, AQURATE and Minibix. By enhancing these toolkits and integrating them with Maxima, a computer algebra system, MathAssess will shown how the specific needs of mathematics can be catered for using open content interoperability standards and open source software. MathAssess acknowledged long-established user needs in mathematics e-assessment, diagnostic, formative and summative providing for the delivery of truly randomized questions and tests, with hints, solutions and feedback being available to students at all appropriate stages.

4.4 SelMa Project

The project “Self-guided Learning in Teaching Mathematics-SEC II", SelMa, is addressed to the mathematics teaching in the High Schools and represents a new web based strategy for learning mathematics. Five authoring schools are working out scenarios, media and materials for phases of self-guided learning, that will be tested systematically by trial schools with regard to their everyday suitability. The core of the project are three different approaches to such learning arrangements: independent learning centre, jigsaw classroom and learning at stations. Learning diaries prove to be useful for the learners' reflection of their learning process. The development as well as the management of such learning arrangements does place new demands on the teachers. The efficacy of this method has been tested in a study investigating the effects of motivational email messages on learner self-efficacy and achievement in an asynchronous college algebra and trigonometry

[18] http://www.imsglobal.org/question/qtiv1p2/imsqti_oviewv1p2.html

[19] www.jisc.ac.uk/

course revealing positive relationship between SELMA and math achievement. This system includes some of the tools abovementioned as general requirements for an online educational activity. It seems limited in the communication between teachers and students, and a conference area is missing while the e-mail communication reduces the possible activities mainly to asynchronous activities.

4.5 MathAssess

MathAssess is a JISC demonstrator project that associated many partners and the most important Universities in United Kingdom. The possibilities offered by this web based platform are different but mainly they let the students attempt questions, get new questions, take tests online while on the other hand the possibility is offered to teachers to design and administer mathematics tests online. The core of this platform is the interaction with synchronous activities that meet most of the requirements abovementioned as essential for successful e-learning online teaching. Expressions are online displayed and manipulated using an easy to use editor and it is possible to create randomized items using templates or compare input and expected answers algebraically. It is possible to use a Computer Algebra System (CAS), compute mathematical values in more complex ways than standard QTI allows, for computing single or groups of mathematical expressions, that can be manipulated and input by the students in a natural way and also authors, e.g. teachers, that can input materials or edit them using MathML. Also CAS allows the comparing of two expressions (variables of base Values) and the evaluating of a condition. Rendering and possibility to represent live graphics and diagrams complete the possibilities of the system.

4.6 ActiveMat

ActiveMat is a research project focused on working in the e-learning of mathematics using a self-regulated learning process. It cooperates with teachers, tutors, psychologists and educators to improve learning and teaching in schools, universities and life-long learning. The research is focused on: adaptive learning environments for mathematics and other fields, collaborative environments, e-portfolios and learning diaries. The project adopts develops and apply techniques from artificial intelligence e.g. semantic knowledge representations, user modeling and planning, educational data mining and natural language processing. The project developed a main platform that includes web based applications user-adaptive for mathematics in school, university and life-long learning, ActiveMath, an interactive, learner-centered platform that can integrate a number of interactive components and tools. This platform has been customized with look and feel, with specific content and with tailored interactivity and tools to various applications like LeActiveMath-Calculus and Standard Activemath with a number of contents including optimization, combinatorial calculus, etc. The system guides the student in a self-regulated learning, can adapt to individual knowledge state and personal interests and learning goals, and allows the learner to realize their learning potential. ActiveMath integrates several functions and tools: computer algebra systems, function plotter, concept map tool, semantic search, notes function. This is supported by an intuitive layout and navigation guidance appropriate for student's needs. It is worth to note also that this platform, developed by a team of pedagogues, is investigating and documenting best practice cases for its use in classrooms in European schools and also the students can engage themselves in the development of contents for ActiveMath, taking part in the building of a next-generation learning system.

4.7 *EMILeA-stat*

EMILeA-stat is a government supported project developed in Germany that focuses on develop and provide a multimedia, web-based, and interactive learning and teaching environment in applied statistics. The project was set up by 13 partners working at eight German universities: Augsburg, Bonn, Berlin (Humboldt-University), Dortmund, Karlsruhe, Münster, Aachen, and Potsdam. The project is also supported by further partners in advice and it cooperates with economic partners such as SPSS Software[20], Bertelsmann Springer Science Business Media (Springer Verlag)[21], and MD*Tech Method & Data Technologies (XploRe-Software). EMILeA-stat, realized a proprietary learning environment and its content, which is also a registered brand name, and has been developed as a system suitable for teaching statistics at schools and universities. Its design offers on the one hand the opportunity to tailor individual courses covering specific learning needs and on the other hand EMILeA-stat serves as an environment for self-directed learning as well as an intelligent statistical encyclopedia. The support of learning and teaching by offering interactive visualizations, such as Java Applets, is another main concept of EMILeA-stat.

4.8 *Webmath*

Webmath is a math-help web site that generates answers to specific mathematics questions and problems, as entered by a user. The answers are generated and displayed real-time, at the moment a web user types in their math problem and clicks "solve." In addition to the answers, Webmath also shows the student how to obtain the right answer. The website is designed using an intuitive layout and it is divided in sections: math for everyone, general math, algebra, plots and geometry, trigonometry etc., that can be interactively used. The user is guided to the answer displayed in an automated self learning environment. It can be used by teachers and by students that need to explore or improve their skills in basic and some advanced mathematics problems. The site is realized using an html editor with applets and a familiar layout that makes it easy to use. This web learning system includes however only some of the above mentioned tools that are considered a general requirements for an online educational activity seeming then to be better addressed as a web learning support for the users compared to a web learning environment. The communication e.g. a forum area or a conference area are missing and e-mail communication reduces the possible activities mainly to the asynchronous ones.

---

[20] www.spss.com/

[21] www.springer-sbm.de/

5 An Integrated Approach of Mathematics in Scientific Education

It is now well established that an up to date teaching approach to scientific disciplines needs mathematics as a basic tool for analyzing and solving real problems (De Corte et al., 2000; Greer, 1993; Greer, 1997; Houstis, 2003; Landau, 1995; Marchioro 1995, 1997). The role of mathematics is not just limited to logical reasoning and quantitative calculation but it is rather considered essential for a thorough understanding of important concepts in applied disciplines and for a deep comprehension of natural phenomena. This strongly requires a reformulation of mathematics education, especially for undergraduates, both in terms of curricula and educational approaches. Moreover, a central role in preparatory math for scientific disciplines is the effective use of up to date available computing tools for the approach to solving applied problems. This is especially true in Engineering education (Klingbeil et al., 2008), where an *inadequate mathematical skills at the freshman level might reduce the competitiveness within the global economy* (Manseur et al. 2009).

However, quantitative thinking and mathematical modeling approaches are becoming dominating in all scientific disciplines, at graduate and undergraduate levels and in high school too, so that Mathematics courses become "service" courses with a very critical role in the overall educational process (Bialek and Botstein, 2004; Brent, 2004; Gross, 2004; Pursell, 2009; Witten, 2005).

Mathematics needs to be integrated in engineering and natural science educational programs according to a problem solving approach that allows:

- to analyze and understand scientific phenomena by using models and simulations;
- to make experimental data analysis and systems simulation;
- to use numerical and computational tools.

In this process computational modeling, numerical analysis, and computing strategies are not the focus of the lessons, but naturally enter as elements of the problem solving process (De Corte et al., 2000; Poggio and Smale, 2003; Marchioro and Landon, 1997).

Such an innovative approach needs tools and techniques which are quite different from the classical ones: several different components are involved in applied discipline lessons to make effective an integrated educational process, e.g.:

1. traditional "learning tools", mainly textbooks and notes;
2. traditional "computational tools" (numerical methods, algorithms, software) available in books and on the net, with technical descriptions and documentations thought mostly for trained users;
3. innovative ICT tools, such as databases, web-learning tools, problem solving environments (PSE) (Avgeriou et al., 2001; Dede, 1996; Eastmond and Ziegahn, 1996), that are going either to replace or to integrate the traditional computational tools.

A critical issue in mathematics for undergraduate and high school students is the need to be efficiently integrated within different courses having different requests in terms of mathematics knowledge. Therefore inter disciplinary aspects, flexibility and adaptability to different educational contexts should be key features of a modern web-learning system for mathematics. The platforms we presented in the previous section, which are very powerful in terms of content management, assessment and interactivity, are still not completely satisfying, although their potentialities in terms of interdisciplinary educational paths are still not fully realized.

In the following pages we present a short outline of SIRMM (Searchable Information Repository of Mathematical Models), which is an attempt to supply an environment for mathematical teaching and learning within scientific disciplines (Giannino et al., 2004a-b). SIRMM has been originally projected at the University of Naples Federico II in the framework of an Italian National Research Project[22]. Currently it is under further development and implementation although its architecture has been mostly defined and a prototype version is running.

5.1 SIRMM Conceptual Architecture

SIRMM proposes a conceptual framework for the development of a problem solving-oriented learning system for scientific disciplines in which the main mathematical concepts involved in real world applications are supplied in a unified, flexible and collaborative framework. It realizes a link between mathematics and real world, providing easy access to indispensable mathematical knowledge. It can be used at different levels of depth and with different paths tailored on students capabilities and needs. Finally, it can be integrated with external resources.

General goals of SIRMM are:

- to stimulate the development of analytic, synthesis, modeling and evaluation capabilities for a better understanding of reality and the creation of possible scenarios;
- to overcome knowledge separation actually existing among different disciplines for a multidisciplinary approach to mathematics;
- to foster the exchange between knowledge producers and consumers, in a process of constant feedback on ideas, contents and methodologies.

SIRMM database contains different kind of Learning Objects (LO) usable in learning processes within different educational paths. It implements a learn-by-doing approach. In contrast to traditional science, students are introduced to a Learning Environment (LE) where they have the option to dynamically create their own study path rather than follow a prefixed one.

SIRMM allows people to solve problems step by step from the easiest level to the most difficult ones. The system provides the possibility of building various modules of information related to every specific issue, and different educational paths. Figure 1 presents two alternative paths: the first one focuses on mathemati-

[22] "Database Multidisciplinare Interattivo", Progetto Formazione a distanza per la integrazione dei saperi nelle scienze della vita, PON 2000-2006.

cal issues (calculus and numerical analysis and computational techniques) integrating application problems in the learning path, while in the second one the mathematical concepts support the formalization and the problem solving stage in a technical discipline.

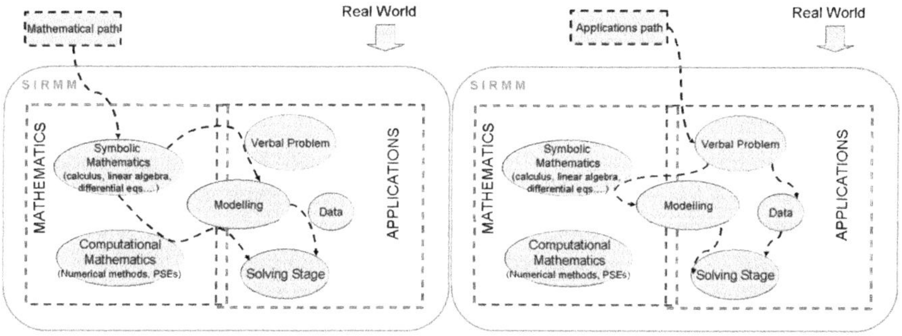

Fig. 1. Example of learning paths supported by SIRMM

The information granularity in SIRMM replaces the traditional structure of knowledge arranged in a rigid form overcoming knowledge separation and obstacles existing in any multidisciplinary approaches. Fine-grained modules are reusable for different courses and disciplines, facing the needs of an educator to teach the desired course and convert a module already developed into another one. SIRMM is flexible enough to address changing user requirements and knowledge, so to be used in different contexts.

SIRMM might also allow to realize mathematics courses based on the problem solving paradigm. Rather than presenting the various arguments one after the other in a linear sequence, a modern cognitive approach suggests to illustrate concepts moving from practical applications and new mathematical concepts are motivated from the need of solving/describing new physical problems coming out in a learning path stimulated by the study of the real world. According to the constructivists theories (Dick, 1992; Hannafin, et al., 1999; Jonassen and Grabowski, 1993, Vrasidas, 2000) new knowledge is added to existing one through a constructive process in which the learner has an active role in constant interaction with the surrounding LE. This is especially appropriate in dealing with mathematics, too often perceived as a mere collection of formulas.

5.2 SIRMM Platform

SIRMM is based on an interdisciplinary interactive and multimedia database. It is accessible through a web browser (http://www.sirmm.unina.it) that integrates different scientific oriented applications. It provides, for each problem:

1. a clear description of the phenomena;
2. a model definition (data, relations, parameters) with the analysis of the mathematical issues;

3. a numerical-computational approach;
4. case studies with analysis and interpretation of the results;
5. links to additional source of information.

5.3 Model Database

SIRMM is composed of a set of entities interacting at different levels. In this section a model for those entities and their relationships is given. The design of the learning management reflects this model. Figure 2 provides an Entity-Relationship model. The entity **DISCIPLINE** describes the scientific area of interest, that is the fields in which the contents of SIRMM are catalogued. Each discipline includes several **PROBLEMs** that are identified by a name and a short description. A problem is formulated as a **MODEL** (possibly more than one) composed, according to Von Neumann, of equations and a description in natural language of what it represents in that context. Each **MODEL** is identified by its name and can be related to different **PROBLEMs**. The following step concerns the computational solving stage which involves a **SOFTWARE** implementing a suitable solving method for the **MODEL** and input **DATA**. Each **SOFTWARE** can be either a code (specifically created for SIRMM) or a link to some external PSE or library. Input **DATA** are given in a format suitable for related **SOFTWARE**. A specific instance of a problem can be described in natural language and becomes a **CASE STUDY**. Each of those entities is related to one or more **RESOURCEs** (e.g. book, article, etc.) to provide further reading on the topic.

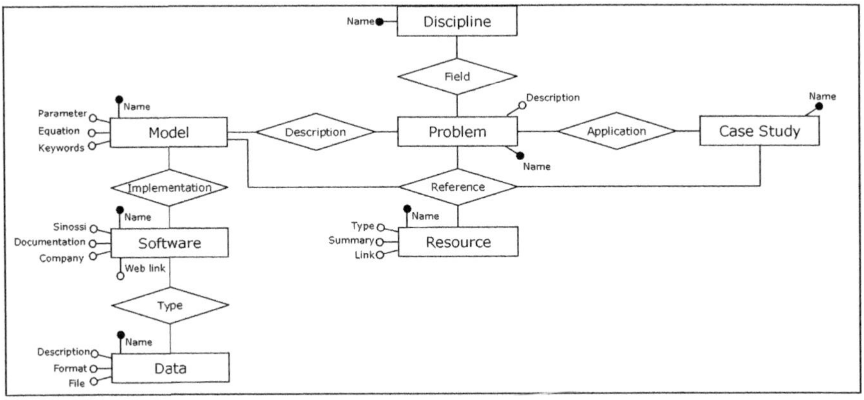

Fig. 2. Entity-Relationship diagram

The logical model of entities and relations induces a model in the navigation of the information stored in the system; since each entity is in relation with one or more other different entities, it is possible to start the navigation from each of them.

5.4 SIRMM Technological Architecture

The technological architecture of SIRMM has been developed using the logical model of layers. In this model, system functionalities are divided and implemented in software elements called components. Each component uses the functionalities of those laying in a lower layer and it is used by at least one component in the upper layer. This approach represents a de facto standard in open learning management systems (Avgeriou, 2001b) which allow implementation modularity, transparent use of lower functionalities, and easy detection of errors or weaknesses. All components are logically divided in the following four layers (Giannino et al., 2004):

- System software layer – It offers components managing basic services provided by operating system, which are not in its kernel;
- Middleware – It contains all system services that are not readily available in the operating system of the server hosting the repository and that are used to deliver it on the internet;
- Application general – It comprises all components that are not specific to SIRMM and that can be re-used by future applications with similar demands;
- Application specific – It encloses all software components that have been specifically implemented for the application and that can be used by other systems with the same needs.

5.5 Using the System

SIRMM includes three types of users with different authorization levels: *consumer/student*, *producer/teacher* and *administrator*. The consumer is a user that can access the database contents without modifying them. The *producer* is a user that can add information in the SIRMM database; usually he is a teacher who wants to insert and share Learning Objects and therefore needs to be authenticated from the system. He can modify only the LO that he has previously inserted. In addition, using the data of SIRMM the producer can build his own e-learning course. Finally, the *administrator* is in charge of all system management tasks such as the authentication requests.

This structure is implemented into SIRMM through three main sections accessible through a full screen menu interface: *Navigate*, *Contribute* and *Course*; any LO in SIRMM system can be accessed by the consumer (*Navigate* section), inserted and modified by the producer (*Contribute* Section). He can also produce an e-learning course, through a logical path which includes LO belonging either to SIRMM or to the external world (*Course* section). Figure 3 shows the flow of information and activities into SIRMM, with the roles of the different kind of users.

To realize a course, once all the elements/pages have been inserted, the producer is able to build and define his course through a system of links to the activity list. These links can be sourced from SIRMM data base elements but also from "external" web resource. In order to exploit the availability of already existing material, the SIRMM course application allows to include any kind of external elements, such as resources available on the word wide web or any material available

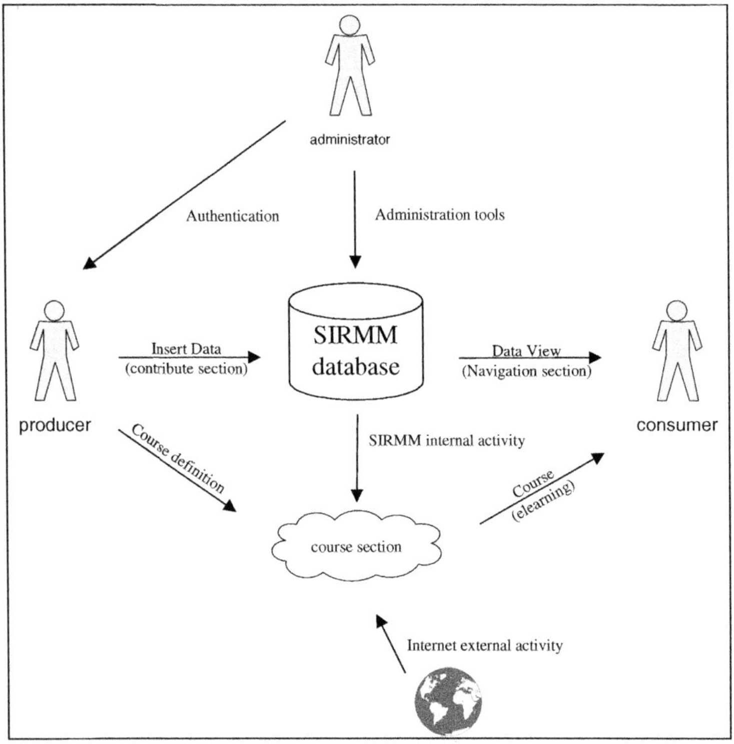

Fig. 3. SIRMM using interfacing structure

in electronic format, or simple links to interesting references. Modularity and Flexibility of SIRMM structure support the creation of independent and well differentiated learning paths based on learning objects (LO) belonging to the system. Moreover, flexibility leads to high level of reusability of the in SIRMM, as well as of entire learning path (courses) that can be easily modified.

A way to exploit the modularity of SIRMM, is to implement a course by re-using units of another course; completely different educational paths can be mapped on large sets of common forms and data. A different view of modularity is to offer the possibility to implement different paths through the material matching the instructional approach to the needs and interests of every student; this is what is called "differentiated instruction"[23], an instructional theory that allows teachers to take diverse student factors into account when planning and delivering instruction. Based on this theory, teachers can structure learning environments that address the variety of learning styles, interests, and abilities found within a classroom. Differentiating instruction is actually an essential tool for integrating technology into classroom activities in an effective way.

Of course the reusability is a feature that depends on the size (granularity) of the learning objects. The lower is the granularity of the contents, the higher is their

[23] http://www.glencoe.com/sec/teachingtoday/subject/di_meeting.phtml

potential for reuse in multiple applications (Wiley et al., 2000) and therefore the potentiality of SIRMM strongly depends on how learning objects will be supplied to the system, e.g. on the way the suppliers of the information (users) will feed the system.

6 Conclusions

Web tools and learning environments available nowadays on the Internet allow to build and share different meta cognitive frames that represent a powerful device in mimicking the learning process and the problem solving experimental approach that is used in all fields of the science, and, specifically, in mathematics. Most of the technical problems connected with the mathematics discipline, related to the specific language have been solved by the new platforms and tools that integrate easy to use interface with powerful semantic oriented editors and problem solving environments. They allow to build learning environment able to semantically search for information when needed, to process them in a meaningful way, to interconnect with web tools and to make contents fully available. However, if hardware and software potentialities are enormous, still much can be done in terms of learning and teaching models to implement.

New educational paradigms are suggested from the new web-leaning environments: mathematical knowledge integration in all scientific disciplines, according to quantitative thinking and modeling approaches, is today really possible through the effective use of technologically advanced web learning environments.

These considerations lead to the necessity to make available platforms and learning environments that, making use of the ICT systems available, produce an interactive and flexible learning-teaching framework able to support contest-tailored synchronous and asynchronous online activities. The SIRMM project, that we described in the last section of the chapter is a possible attempt to do this through a problem solving-oriented learning system for scientific disciplines in which the main mathematical concepts are supplied in a unified, flexible and collaborative framework. We hope that similar and more effective learning environments will be soon made available so to foster a modern and effective mathematics education.

References

Avgeriou, P., Papasalouros, A., Retalis, S.: Learning Technology Systems: issues, trends, challenges. In: Proc. of the 1st Int. Organization for Science and Technology Education Symposium in Southern Europe: Preparing future citizens (2001a)

Avgeriou, P., Retalis, S., Skordalakis, M.: An Architecture for Open Learning Management Systems. In: Panhellenic Conference on Informatics, pp. 183–200 (2001b)

Bialek, W., Botstein, D.: Introductory Science and Mathematics Education for 21st-Century Biologists. Science 303(5659), 788–790 (2004)

Brent, R.: Points of View: The Interface of Mathematics and Biology: Intuition and Innumeracy. CBE Life Sci. Educ. 3, 88–90 (2004)

De Corte, E., Verschaffel, L., Greer, B.: Connecting mathematics problem solving to the real world. In: Proceedings of the International Conference on Mathematics Education into the 21st Century: Mathematics for Living, pp. 66–73 (2000)

Dede, C.: The evolution of distance education: Emerging technologies and distributed learning. The American Journal of Distance Education 10(2), 4–36 (1996)

Dick., W.: An instructional designer's view of constructivism. In: Duffy, T., Jonassen, D. (eds.) Constructivism and the technology of instruction: A conversation, Hillsdale (1992)

Driscoll, M.: Psycology of Learning for Instruction. Needham Heights/Allyn & Bacon, MA/New York (2000)

Eastmond, D., Ziegahn, L.: Instructional design for the online classroom. In: Berge, Z.L., Collins, M.P. (eds.) Computer mediated communication and the online classroom, pp. 59–80. Hampton Press, Inc., Cresskill (1996)

Giannino, F., Guarracino, M.R., Monetti, V.M., Romano, N., Toraldo, G.: SIRMM: Searchable Information Repository of Mathematical Models. In: EISTA 2004, International Conference on Education and Information Systems: Technologies and Applications, Orlando, Florida, USA (2004a)

Giannino, F., Guarracino, M.R., Monetti, V.M., Romano, N., Toraldo, G.: An Adaptable Learning Technology System for Mathematical Models. In: ICALT 2004, The 4th IEEE International Conference of Advanced Learning Technologies, Joensuu, Finland (2004b)

Graziadei, W.D., Gallagher, S., Brown, R.N., Sasiadek, J.: Building Asynchronous and Synchronous Teaching-Learning Environments: Exploring a Course/Classroom Management System Solution (1997), http://horizon.unc.edu/projects/monograph/CD/Technological_Tools/Graziadei.asp

Greer, B.: The modelling perspective on the wor(l)d problems. Journal of Mathematical Behavior 12, 239–250 (1993)

Greer, B.: Modelling reality in mathematics classroom: the case of word problems. Learning and instruction 7, 293–307 (1997)

Gross, J.L.: Points of View: The Interface of Mathematics and Biology: Interdisciplinarity and the Undergraduate Biology Curriculum: Finding a Balance. Cell. Biol. Educ. 3, 85–87 (2004)

Hannafin, M., Land, S., Oliver, K.: Open learning environments: Foundations, methods, and models. In: Reigeluth, C.M. (ed.) Instructional design theories and models: A new paradigm of instructional theory, vol. II, pp. 115–142. Lawrence Erlbaum Associates, New Jersey (1999)

Harasim, L.: Online education: The future. In: Harrison, T.M., Stephen, T. (eds.) Computer networking and scholarly communication in the twenty-first-century university, pp. 203–214. State University of New York Press, New York (1996)

Houstis, E.N.: The Role of Problem Solving Environments in Engineering and Mathematics Education. In: Proceedings of the 6th International Conference on Technology in Mathematics Teaching. Telematics Network Centre, University of Thessaly (2003)

Klingbeil, N.W., Rattan, K.S., Raymer, M.L., Reynolds, D.B., Mercer, R.E.: A National Model for Engineering Mathematics Education. In: NSF CCLI PI Conference, Washington, DC (August 2008)

Knuth, D.E.: The TeXbook. Addison-Wesley, Reading (1984)

Jonassen, D.H., Grabowski, B.L.: Handbook of individual differences, learning and instruction. Lawrence Erlbaum Associates, Hillsdale (1993)

Keegan, D.: Foundations of Distance Education. Routledge, New York (1996)

Kong, S.C., Kwok, L.F.: An interactive teaching and learning environment for graph sketching. Computers & Education 32, 1–17 (1998)

Landau, R.: Computation in Modern Physics. IEEE Computational Science and Engineering 2(3), 84–85 (1995)

Manseur, R., Ieta, A., Manseur, Z.: Reforming Mathematics Requirements For A Modern Engineering Education. In: 39th ASEE/IEEE Frontiers in Education Conference (2009)

Marchioro II, T., Martin, D., Payne, W.: UCES: An Undergraduate CSE Initiative. IEEE Computational Science and Engineering 2(3), 69–73 (1995)

Marchioro II, T.L., Landon, R.H.: Web-Based Education in Computational Science and Engineering. IEEE Computational Science & Eng. 4, 19–25 (1997)

Michailidou, A., Economides, A.: Elearn: Towards a Collaborative Educational Virtual Environment. Journal of Information Technology Education 2, 131–152 (2003)

Moallem, M.: Applying Constructivist and Objectivist Learning Theories in the Design of a Web-Based Course: Implications for Practice. Educational Technology & Society 4(3) (2001)

Nagy, A.: The Impact of E-Learning. In: Bruck, P.A., Buchholz, A., Karssen, Z., Zerfass, A. (eds.) E-Content: Technologies and Perspectives for the European Market. Springer, Berlin (2005)

Nichols, M.: E-Learning in context (2008)
`http://akoaotearoa.ac.nz/sites/default/files/ng/group-661/n877-1-e-learning-in-context.pdf`

Oppermann, R., Rashev, R., Kinshuk: Adaptability and Adaptivity in Learning Systems. In: Behrooz, A. (ed.) Knowledge Transfer VII, pp. 173–179. pAce, London (1997)

Pahl, C., Holohan, E.: Applications of semantic Web technology to support learning content development. Interdisciplinary Journal of E-Learning and Learning Objects 5, 1–25 (2009)

Poggio, T., Smale, S.: The Mathematics of Learning: Dealing with Data. Notices of the American Mathematical Society (AMS) 50(5), 537–544 (2003)

Pohjolainen, S., Hantakanges, S., Ranta, P., Levasma, J., Pesonen, K.: A learning experiment in mathematics using A&O-learning environment. Int. J. Cont. Engineering Education and Lifelong Learning 13(1/2) (2003)

Pursell, D.P.: Enhancing Interdisciplinary, Mathematics, and Physical Science in an Undergraduate Life Science Program through Physical Chemistry. CBE Life Sci. Educ. 8, 15–28 (2009)

Rafaeli, S., Barak, M., Dan-Gur, Y., Toch, E.: QSIA – a Web-based environment for learning, assessing and knowledge sharing in communities. Computers & Education 43, 273–289 (2004)

Rekkedal, T., Qvist-Eriksen, V.: Internet Based E-learning, Pedagogy and Support Systems. NKI Distance Education (March 2003)

Salmon, G.: E-moderating: The Key to Teaching and Learning Online. Open and Distance Learning Series. Kogan Page Ed., London (2000)

Siemens, G.: The Whole Picture of E-learning (2003),
`http://www.elearnspace.org/Articles/wholepicture.htm`

Sheremetov, L., Arenas, A.G.: EVA: an interactive Web-based collaborative learning environment. Computers & Education 39, 161–182 (2002)

Smith, P.L., Ragan, T.J.: Instructional design. Prentice-Hall, Upper Saddle River (1993)

Vrasidas, C.: Constructivism versus objectivism: Implications for interaction, course design, and evaluation in distance education. International Journal of Educational Telecommunications 6(4), 339–362 (2000)

Wiley, D.A., Recker, M., Gibbons, A.: Getting axiomatic about learning objects (2000), http://reusability.org/axiomatic.pdf (retrieved from the Internet on May 5, 2004)

Witten, G.Q.: Designing a Mathematics Course for Chemistry and Geology Students. Educational Studies in Mathematics 58(1), 1–19 (2005)

Author Index

Batch number: 10755292

Printed by Printforce, the Netherlands